D0848117

PAYING FOR ENERGY

Report of

The Twentieth Century Fund Task Force
on the International Oil Crisis

Background paper by
SIDNEY S. ALEXANDER

McGraw-Hill Book Company

New York   St. Louis   San Francisco   London   Düsseldorf
Kuala Lumpur   Mexico   Montreal   Panama   São Paulo
Sydney   Toronto   New Delhi   Singapore

THE TWENTIETH CENTURY FUND is a research foundation engaged in policy-oriented studies of economic, political, and social issues and institutions. It was founded in 1919 by Edward A. Filene, who made a series of gifts that now constitute the fund's assets.

Library of Congress Cataloging in Publication Data

Twentieth Century Fund. Task Force on the International
    Oil Crisis.
    Paying for energy.

    Includes bibliographical references.
    1. Petroleum products—Prices—United States.
2. Petroleum industry and trade—Finance.   3. Price
policy.   I. Title.
HD9564.5.T85   1975   338.2'3   75-38644

ISBN 0-07-065621-5
ISBN 0-07-065622-3 pbk.

338.23
T971

Contents

Foreword

The economic and psychological shock waves caused by the oil embargo of 1973 and the doubling of the price of oil that immediately followed are still reverberating. Although not all the fears that were first voiced have materialized, the events of 1973 brought about profound and significant changes in international financial and political relationships, changes that in many respects are still to be felt. There is a general recognition that the United States can never return to the old days, when supplies of energy were abundant and cheap and the nation was profligate in its consumption habits. But there also is considerable confusion about what policies ought to be pursued in this new and challenging situation. In 1974, regarding the financial problems wrought by the energy crisis as particularly critical and urgent, the Trustees of the Twentieth Century Fund approved the establishment of an independent Task Force to deliberate over and report on them.

This project was not the first manifestation of the Fund's concern over energy policy. Early in the last decade, the Fund published a study calling for the development of a new civilian nuclear-energy policy as an alternative to overdependence on fossil fuels. In 1971, well before the energy crisis hit the headlines, the Fund sponsored another project dealing with the need for a national energy policy that would take into account the need for conservation of petroleum and for innovation in exploring and exploiting new and cleaner sources of supply. In

v

fact, for much of the last twenty years, the nation and the rest of the industrialized world have been drifting toward an energy crisis. Throughout this period, too little heed was paid to the few who warned that in an age of energy the lack of effective policies to assure adequate supplies at an economic price was an invitation to disaster.

The members of the Task Force assembled to deal with the energy crisis were diverse and distinguished. The group included oil experts who had been prominent for sounding the alarm about energy at an early date, as well as economists, bankers, and diplomats who had long been involved in the problems of the international monetary network. That network, of course, has had to bear the initial impact of the sustained increase in the price of oil. So it was both appropriate and useful that the Task Force focused its considerable talent and efforts on the financial and economic aspects of the crisis.

Because the group was diverse, it is not surprising that its members frequently disagreed, politely and firmly, with one another. In the course of their long and intense discussions, they did not arrive at any clear and simple solutions for dealing with the crisis. On the contrary, their Report serves to demonstrate that there are no simple or clear solutions. But what they have succeeded in doing is to illuminate and clarify the complicated issues and explain their relationship. The result is a document that will help to make possible more informed and pointed debate about the policy options that are available.

The Fund is grateful to the entire Task Force for its dedicated labors. But I must pay special tribute to Chauncey E. Schmidt, its chairman, who carried out, with great patience, tact, and diplomacy, the difficult assignment of bringing about a meeting of minds on many issues, and to Sidney Alexander, who served as the Task Force's rapporteur and wrote the informative backgound paper that accompanies the Report. If the Task Force does not have all the answers—and nobody does—it asked all the right questions. The light it sheds on the energy crisis and what should be done about it represents a significant public service.

M. J. Rossant, *Director*
The Twentieth Century Fund
December 1975

Members of the Task Force

M. A. Adelman
Professor of Economics,
Massachusetts Institute of
Technology
Cambridge, Massachusetts

Robert O. Anderson
Chairman of the Board,
Atlantic Richfield Company
Los Angeles, California

Edward M. Bernstein
President, EMB Limited
Washington, D.C.

Benjamin V. Cohen
Lawyer
Washington, D.C.

Peter M. Flanigan
Managing Director, Dillon
Read & Co., Inc.
New York, New York

Charles J. Hitch
President, Resources for the
Future, Inc.
Washington, D.C.

Peter B. Kenen
Walker Professor of
Economics and
International Finance,
Princeton University
Princeton, New Jersey

Walter J. Levy
Consultant, W. J. Levy
Consultants Corp.
New York, New York

John R. Meyer
1907 Professor in
Transportation, Logistics,
and Distribution, Harvard
Business School
Boston, Massachusetts

Leslie C. Peacock
Vice-Chairman of the Board,
Texas Commerce
Bancshares, Inc.
Houston, Texas

John S. Samuels III
Chairman of the Board,
Carbomin International
Corporation
New York, New York

Nathaniel Samuels
Chairman of the Board, Louis Dreyfus Corporation
New York, New York

Chauncey E. Schmidt,
Chairman
President, The First National Bank of Chicago
Chicago, Illinois

Allan Sproul
Former President, Federal Reserve Bank of New York
Consultant, Wells Fargo Bank, N.A.
San Francisco, California

Herbert Stein
A. Willis Robertson Professor of Economics, University of Virginia
Charlottesville, Virginia

Richard J. Whalen
Author
Fort Sumner, Maryland

Sidney S. Alexander,
Rapporteur
Professor of Economics, Massachusetts Institute of Technology
Cambridge, Massachusetts

Report of the Task Force

The discovery of the nations that make up the Organization of Petroleum Exporting Countries (OPEC) that they could dictate the price of crude oil to the rest of the world ranks as one of the most significant events of the post-World War II era. But today, only two years after OPEC unilaterally imposed a fivefold price increase, complacency about the future cost and availability of oil is widespread in the industrialized West. Initially, of course, OPEC's move engendered apocalyptic visions among the oil-importing nations. But these visions have failed to materialize, and the world recession, the worst general economic decline since the Great Depression, has masked the effects of OPEC's actions. In fact, the increase in the price of oil not only has intensified the recession around the world but also is acting to slow the recovery that is now under way. It also has distorted international trade and finance and hampered the economic prospects of the less-developed world in ways not yet fully realized. This report is addressed to the legacy of OPEC's action and the burden it has placed on the future well-being of the world economy.

The problems posed by the oil cartel's price rise remain serious and potentially explosive, even though they seem a far cry from those predicted at the end of 1973 when for the first time OPEC raised its prices without negotiating with the oil companies. Initially, concern in the West focused on three main worries:

1. That some of the oil-exporting nations would be earning much more than they could immediately spend or invest and would therefore choose to cut production and raise prices further. Some authorities warned that

3

the West would have to provide OPEC with attractive opportunities for investing its surplus earnings, or "petrodollars," so that they could flow back in payment for uninterrupted production of oil and thus resolve what has been called "the fundamental problem of primary recycling."

2.   That when this huge flow of petrodollars was "recycled" between OPEC and the oil importers through the international financial system, that system would come under unmanageable strains. These strains would comprise what are termed the "special problems of primary recycling."

3.   That even if primary recycling were successful, some nations would have more nearly adequate access than others to OPEC funds. Uneven distribution of OPEC funds would create "secondary recycling" problems. Some nations slighted by OPEC might have to resort to "begger-thy-neighbor" policies in order to win relief from huge payments deficits. The result of these policies, according to advocates of this line of argument, would be a reversal of the postwar progress toward freer and expanding world trade.

The concept of recycling is essential to an understanding of much that follows in this Report. To begin with, oil-importing nations can pay for their petroleum in two basic ways: by selling their own goods and services to the oil exporters, or by transferring claims. Such transfers may involve going into debt, selling assets, or shifting reserves. Nations have been using all three forms of transfer, and over the next few years, the accumulated amount may reach dangerously high levels. Until now, the individual consumer has borne the real and direct burden of the oil price increase, paying more in exchange for gasoline, heating oil, and the many other products of OPEC crude. The added burden on consumers has also contributed to both inflation and recession. Yet the full cost of imported oil has not yet come home to the importing countries, because, to date, less than half of the money and other claims transferred to the oil producers has been redeemed for actual goods and services.

We call it "primary recycling" when the OPEC nations

accept transfers of claims in payment for their oil. Most of the claims are in the form of the customary financial instruments—bank deposits, short-term government paper, accounts receivable, public and private bonds or other certificates of indebtedness. These claims on future cash remain in the hands of intermediaries (e.g., banks), and these intermediaries may encounter a wide variety of special problems in handling these funds. For example, OPEC may wish to deposit more money on demand than banks can safely place in long-term loans. Apart from such problems, OPEC nations obviously do not place their funds with intermediaries in non-OPEC nations in precise proportion to each non-OPEC nation's oil-trade balance with OPEC. Importing nations that come up short then have to borrow, in turn, from the intermediary financial and asset markets, and this borrowing is called "secondary recycling." The less creditworthy such borrows may be, the less able will they be to market their debt. A lot of OPEC money is placed in the Eurocurrency market, which relends it in turn but is becoming increasingly sensitive to the quality of a borrower-nation's credit.

In the past two years, a great deal of recycling has been accompanied by a great transfer of wealth. If the increase in price from 1971 to 1975 were multiplied by the volume of oil imports in 1973, it would total over $90 billion, or 2 to 3 percent of the national incomes of the importing countries. In fact, the increase in the cost of oil imports in 1974 was some 12 percent less than that because of slowing demand (as prices rose, people bought less), the economic recession experienced in varying degrees around the world, and the generally mild winter. Of the $100 billion spent on oil imports in 1974, $70 billion was paid in transfers of claims.[1]

Despite the magnitude of these sums, the worst of the initial fears has failed to materialize: OPEC has kept on producing, in exchange for a continuing flow of claims as well as rapidly rising imports of goods and services; indeed, the sellers have been anxious to sell; the finanical markets have largely overcome their spcial problems so far; and the major importing nations have managed their international borrowings, payments, and accounts. The deep pessimism of 1973 and 1974 has given way to complacency in 1975, apparently justified by certain developments.

First, consumer demand has fallen, forcing the OPEC countries to cut their production drastically below capacity in order to maintain the price. At the peak of the production cutbacks, more than 30 percent of the world oil industry's capacity to produce was going unused. Even with a strong sales recovery in 1976, this excess capacity will probably remain over 20 percent of the total developed capacity, which itself is far below the potential of the fields already discovered. This excess producing capacity reduces the likelihood that the oil-exporting countries, as a group, will cut back production because of the lack of attractive foreign assets. The populous oil producers without a cash surplus will want to increase their production in order to increase revenue. Some have already shown considerable concern with the loss of market shares. At the September 1975 OPEC meeting, some less-favored countries demonstrated their sensitivity to the various quality and freight discounts and premiums which have tended to diminish their sales. The sparsely populated producers with a surplus might indeed be able to afford additional cuts in production, but the populous, non-surplus producers would be more than willing to make up the difference. Even for the surplus producers, oil in the ground is not necessarily worth more than money in the bank; the price of oil in 10 years is not likely to exceed the sum of the current price and the return on investments that can be made with it in the coming decade.* In fact, it would be in the economic interest of each OPEC country to sell as much as it could at the current price, without undercutting the cartel's price level.

Second, the payments surplus in the OPEC countries has begun to decline, thanks to the declining volume of sales and an astonishing expansion of OPEC imports. In 1975, the current-account surplus fell to about $40 billion.

---

*Mr. Adelman comments:
    Even for the surplus producers, it is rare and improbable for oil in the ground to be worth anything like money in the bank. If the opportunity cost of producing a barrel of oil today is its nonproduction in 25 or more years, its present value is roughly a tenth at a reasonable discount rate.

Third, even though the international financial system was undergoing a period of stress and credit tightening, largely for separate reasons, it has handled the OPEC claims transfers without especial difficulty.

And fourth, current-account deficits have been much less of an immediate problem than anticipated, at least for the industrialized countries. In May 1973, the 24 members of the Organization for Economic Cooperation and Development (OECD) showed their willingness to avoid the most injurious of trade policies by taking a pledge that they renewed for the most part a year later (although the pledge does allow a range of actions that could still have the consequence of exporting recession). Italy, the cause of greatest worry among the developed countries, achieved a remarkable turnaround in its trade balance and eliminated its non-oil deficit, although at the expense of a serious recession which left its political future still clouded. Great Britain still runs an unprecedented peacetime deficit, but has managed to finance it through recycling, in large part from OPEC countries. Even the so-called Fourth World, the non-oil-exporting countries of Asia and Africa, has in aggregate managed to finance its huge deficits, in large part attributable to the oil price rise, and even managed a 12 percent increase in real imports in 1974.

Relief is an understandable reaction to these developments, but the newly prevailing optimism, the Task Force believes, underestimates the seriousness, extent, and probable duration of the problems that remain.

Current trends cannot be expected to continue indefinitely. Oil imports will undoubtedly increase, intensifying payment problems, when worldwide economic recovery, welcome as it will be, takes hold. Clogged ports and inadequate infrastructure may dampen OPEC's import boom. Furthermore, some OPEC countries may become disillusioned with their massive investments and development programs, as the inevitable instances of wasted money come to light. The Eurocurrency market may not always be accommodating. These developments are not all equally probable, and none of them is certain, but all must be allowed for.

Moreover, in a temporary fashion which is now coming

back to haunt their own economies, the OECD countries col-
lectively have been able to transform their trade balance from
the deficit of 1974 into a substantial surplus this year, at the
expense of all other nations. Early in 1975, the trade deficits
of non-OECD countries, excluding OPEC, had mounted toward $50 billion annually.[2] The current-account deficit of the
Fourth World group of less-developed nations has mounted
from $9 billion in 1973 to $18 billion in 1974 and to $35
billion in the first part of 1975. Although these deficits result
only in part from oil prices and in part from the impact of the
recession on other raw-material earnings, they are insupport-
able via secondary recycling on any sustained basis. World
trade is now beginning to shrink, reducing the exports of the
industrial nations and depressing their domestic economic
activity. This belated effect of the oil payments problem has
already moderated and will continue to moderate the pattern
of recovery from the world recession. It will severely curtail
the economic growth rates of Fourth World nations, en-
danger their internal stability, and exacerbate their demands
for—and the general need for—worldwide economic policies
that can ease their payments positions and the shrinkage in
world trade.

The oil price rise directly contributed an increase of
about 3.5 percent to the general price level in the United
States and comparable amounts elsewhere before the latest
price increase of October 1975. Furthermore, the increased
cost of oil products diverted consumer expenditures from
domestic products, reducing domestic aggregate demand.
This decline in demand has accentuated the slowdown in
economic activity. Moreover, the leading Western governments failed to recognize the strength of the recessionary
forces, both general and oil-induced, and to appreciate the
special nature of the inflation they were experiencing. They
adopted strong anti-inflationary policies which, in the United
States at least, may have imposed indirect costs, such as added
unemployment and loss of production, which exceed the direct additional cost of the oil imports.

RECOMMENDATIONS

The best remedy for the problems caused by the increased price of oil would be, simply, to lower the price. The Task Force believes that this remedy should be sought through reliance on market forces, *not* through negotiation with OPEC.*

> • *Negotiations on price or commodity indexation agreements must be avoided. They would either be ineffective or serve to keep prices at their present high level.*
> • *Market pressures on the price set by OPEC should be strengthened primarily by enlarging the world's capacity to produce oil and alternate fuels, but also by increased conservation efforts. The best prospects for increasing energy supplies in the next 15 years lie in expanding the production of crude oil and natural gas. Accordingly, the most effective means of exerting market pressure will be to accelerate exploration for crude and develop producing capacity from new sources.*

Until this effort becomes fully effective, the non-OPEC nations must continue to pay attention to recycling surplus oil revenues, both from oil producers to importers and among the importers themselves.

Mr. Cohen wishes to add:
Negotiation with OPEC to lower the price of oil may serve no purpose so long as OPEC exercises its power to fix the price of oil without the concurrence of nations dependent on OPEC crude and is able, by agreement among its members, to limit the supply to the extent necessary to maintain its fixed price.

But that does not mean we should wait indefinitely, until the market pressure of new sources of supply makes it impossible for OPEC to maintain its fixed price unilaterally.

Instead of vaguely threatening the use of force if the price becomes intolerable, we should now urge in appropriate forums the establishment of codes of fair practices to prevent the disruption of international trade; these codes would, among other things, outlaw the manipulation of the price by cartels for their own benefit through their monopolistic or oligarchical control of supply.

• *Non-OPEC countries should welcome OPEC investments as an outlet for the petrodollar surplus. Present United States regulations, freely permitting foreign investment in all but a few areas involving national security, seem to be adequate. The present provisions for government monitoring should suffice in evaluating the continued effectiveness of these regulations. But the government of the United States should not permit Arab governments to coerce American business into abiding by the terms of the Arab economic boycott against Israel, either by undertaking a second boycott against American companies dealing with Israel or by discriminating against Jewish personnel.*

• *Oil-importing countries must remain on guard against beggar-thy-neighbor policies. A general and simultaneous movement to reduce deficits would defeat itself and disrupt world trade.*

Finally, the most severely affected victims of the price rise, the so-called Fourth World, should receive special help.

• *Renewed efforts should be made to increase capital flows, both commercial and official, to these countries. Such efforts should include maintenance of the growth of flows from the developed countries and from the OPEC countries. Although the latter have made projected $10 billion of commitments to the less-developed countries for 1975; the flows have not yet been commensurate with these commitments. OPEC and the developed countries should explore agreements to enlarge aid flows to the developing world. Without such agreements, the OPEC flows cannot be taken for granted.*

• *The United States should continue to pursue trade liberalization for the ·benefit of the Fourth World. The Generalized System of Preferences, designed to lower tariff barriers on manufactures of the less-developed world, should be supplemented by elimination of non-tariff barriers, notably import quotas.*

• *Priority in receiving aid should go to the poorest nations of the Fourth World, which contain over half of the population of this group. Such aid-giving provides the most constructive area for cooperation between the oil importers and OPEC.*

Many of these recommendations run parallel to American policy initiatives outlined in Secretary of State Henry Kissinger's statement to the Seventh Special Session of the UN General Assembly. Although it remains to be seen how Secretary Kissinger's proposals will be translated into practice, the Task Force applauds their emphasis on trade and capital recycling, rather than on commodity cartelization and price negotiations.

BRINGING DOWN THE PRICE

If the price of oil, like that of most other commodities, were set by buying and selling on the marketplace, talk of bringing it down by negotiation or other means would be wishful thinking.* But the world oil market is dominated by the OPEC cartel, which has arbitrarily fixed the price of oil at more than twenty times the cost of production at the most prolific producing area, the Persian Gulf. OPEC's power to fix the price rests ultimately on its own cohesion and the limited ability of non-OPEC producers to increase their own output of oil and alternate fuels.

Although OPEC itself was founded in September 1960, it first acquired its ability to force up the price in the "leapfrogging" negotiations of 1970 and 1971. In September 1970, a new revolutionary government in Libya demanded and won a price increase from its oil companies, ignoring both the economic provisions and the arbitration procedures of the existing concession agreements. In December, the Gulf producers in turn demanded, and received, roughly comparable terms. Thereupon, the Libyans rejected the five-year price schedule they had previously imposed and made further demands. The Gulf producers once more followed suit. The results of these two negotiations were the Teheran agreement

*Mr. Adelman comments:
Indeed, as the example of natural gas price regulation in the United States shows, an attempt to fix the prices below the competitive level dries up supply and inflicts heavy economic costs.
Mr. Meyer concurs.

of February 1971 and the Tripoli agreement of April 1971, which promised five years of stability.

Under the terms of these agreements, prices rose in January 1972 and June 1973, ostensibly to offset dollar devaluations. But in October 1973, OPEC cast these agreements aside announcing that it would henceforth set prices unilaterally.

This record suggests strongly that no negotiations will result in a lower price for oil than OPEC is able to impose on its own. The most effective means to limit OPEC's price-fixing ability lies not in government-to-government talks but in the market pressures of supply and demand. The United States government has sought to get the price of oil down by negotiations with OPEC members, and in spite of the earlier failures, this approach apparently still remains an option in the current unsettled state of American petroleum diplomacy. *But important as it is to obtain a lower price, the Task Force believes that attempts to negotiate with OPEC on price will not serve the interests of the United States and the other oil importing nations.**

Such talks have so far foundered on an astonishing diplomatic development: OPEC, whose actions have hurt the Fourth World more than any other region, has succeeded in retaining political support by channeling the economic grievances of the developing countries into the campaign for a "new international economic order." In the radical presentation of this idea, this new order would be based on raw-material cartels modeled after OPEC. The Paris meeting in April to plan a conference of oil producers and consumers collapsed after Algeria demanded that the agenda include consideration of long-term price agreements for other commodities. Aside from the deficiencies of commodity price agreements in general, we find it unlikely, for political and

---

*Mr. Adelman comments:

OPEC is a collective monopoly composed of sovereign states. The only sanctions which can be brought to bear on any buyer or seller are those of competition and the law. A group like OPEC is beyond either sanction. Hence an agreement with such a group is simply "inoperative" in the true sense of the word.

Mr. Meyer concurs.

economic reasons, that such a conference could significantly lower the price of oil.

Saudi Arabia is the only country with the economic power to break the price. As the "swing man," or residual supplier, Saudi Arabia might do as well in total revenues at $7 per barrel as at $11.51 per barrel (in 1975 dollars), because it would be the principal beneficiary of any increase in total OPEC export volume. But the other producers would suffer the price cut without much increase of volume. In OPEC meetings, Saudi Arabia advocated a lower price publicly but acceded with apparent reluctance to at least modest increases for the sake of maintaining good terms with other producers. But at certain junctures when its unilateral action might have been enough to bring the price down, Saudi Arabia has refrained from attempting any such action. The most notable of these missed opportunities was the Saudi cancellation of an auction of large quantities of crude in August 1974, when the market price showed signs of weakening and might have collapsed altogether under the impact of such a large new supply of oil.

Proposals to "index" the price of oil against an average of the prices of a selected group of commodities and manufactured goods imported by OPEC from the West have been made. Even without an indexation agreement, the OPEC nations may seek to maintain the real price of oil by raising its nominal price to correspond to increases in the general price level; but an indexation agreement would formalize such an arrangement. It also would be subject to a ratchet effect by which prices could be levered up but never down, and it would intensify inflationary tendencies in the general price level.

The basic objection to price negotiations and indexation agreements is that they impede the operation of market forces, which, although they do not work freely in today's world oil market, still can be expected to work for the reduction of the oil price. Oil importers, by increasing their own supply of petroleum and alternate fuels and by controlling their demand for them, do have the potential to exert leverage on several OPEC nations which account for a substantial share of OPEC output. The populous, non-surplus OPEC nations are under heavy pressures to maintain their export level and revenue inflow, even though these pressures may not be felt by the sparsely populated cash-surplus producing countries.

One analyst argues that the members of OPEC could tolerate a surplus capacity of about 45 percent, entailing a production cut to a level of 20 million barrels a day (m.b.d.), before their revenue would fall below their current payment needs.[3] But as OPEC imports approach the level of export earnings from oil and other goods, the readiness of each country to cut production below capacity in order to maintain the cartel price will be reduced; Iran has already been forced to return to the international credit market to finance its development program. But the facts do not justify complacency on this account, because OPEC surplus capacity will be reduced in the next year or two as worldwide economic recovery spurs oil consumption and OPEC production and sales.

Consuming countries can exercise three general sorts of pressure to lower OPEC sales. They may curtail demand by allowing the increased price to exert its usual dampening effect on consumer preferences.* They also may embark on conservation programs, although conservation alone may cut consumption by only 3 m.b.d. or so, even if political obstacles are overcome, and will generate considerable economic costs. Consuming countries also can begin at once to develop alternate energy sources, such as coal, nuclear energy, solar

---

Mr. Stein comments:
An important step the United States can take to reduce imports and put downward pressure on the OPEC price of oil is to remove ceilings on the price of oil produced in this country. Under present regulations, the price of crude oil in the United States is the average of the controlled price for "old" domestic oil and the world price, which we do not control. The lower we keep our price ceiling, the lower this average will be for any given world price, the higher will be our consumption and imports, and the easier it will be for OPEC to maintain a high price. Contrary to the common notion, allowing our own price to rise is a way to get the OPEC price down. Removing price ceilings on domestically produced natural gas also will tend to lower prices of OPEC oil by encouraging more production of an important alternative energy source.

Mr. Meyer concurs.

energy, and synthetic crude derived from coal, shale, or tar sands. These sources may be useful for national security, insofar as they lessen dependence on imported energy, but they also will be at least as expensive as oil is now.*

In addition, consuming nations can accelerate world exploration and development of producing capacity for conventional crude oil itself. Countries that discover large new reserves may wish to join OPEC, but since they will want their share of the world market, every large new discovery, whether inside or outside of OPEC, will add to surplus capacity. *This approach, in our judgment, offers the most promise for the next 15 years. At the present price, market rewards will speed this process, if the many institutional obstacles can be overcome.*

In country after country, the important limits on exploration and development are administrative and legal rather than economic. In the United States, for instance, Alaskan reserves lay untouched for four years after initial development because of the prolonged environmental controversy over the Alaskan pipeline. The Naval Petroleum Reserve No. 4, another potentially large but unexplored area, remains undeveloped because of congressional opposition. Development of offshore oil has been delayed by a jurisdictional dispute between state and federal governments. Many other countries, concerned to protect their national interests, have hindered effective use of their resources.

These concerns are often well-founded, and we do not propose that they be ignored. *But the required safeguards should be formulated as quickly and clearly as possible.* Advanced planning on national policy will shorten the delays and reduce the costs of development. Non-OPEC countries should explore means by which they can conduct the fullest possible exchange of experiences in setting up institutional arrangements and environmental safeguards.

Under a free market, the present high prices would serve as incentives for development. But under current tax arrangements, relatively little of this price will go to those who

---

*A future discussion of domestic energy policy will deal with the national aspects of stimulating production and implementing conservation.

discover the oil and develop the productive capacity. Countries that are trying to capture the benefits of the oil price through taxation will eventually have to support exploration and development by special incentives. The rate at which productive capacity will be built up largely depends on the speed with which governments come to terms with these realities. And productive capacity, both within OPEC and outside it, is the principal instrument for eventually curtailing the power of the cartel.

FINANCING THE SURPLUS: PRIMARY RECYCLING

By 1980, the OPEC countries should have developed to the point at which they spend as much on imports and aid grants as they take in from their oil exports. When they do, the transfer of claims from the oil-importing countries should approximate zero, but the accumulated OPEC surpluses may approach $250 billion. In the meantime, a number of problems have been predicted both for primary recycling, the flow of these funds between OPEC and the rest of the world, and secondary recycling, the flow among the non-OPEC countries.

A number of experts had feared that the OPEC countries, lacking attractive investments for their surplus revenue, would prefer to sell only enough oil to meet their own budget needs and would leave the oil demands of the industrial world unsatisfied. But in fact, the OPEC countries' selling strategy has focused simply on the price of oil. After each government has determined what its take per barrel should be, in taxes or direct sales price (in the case of a government-owned company), it has left sales volume to the vagaries of market demand and the salesmanship of concession-holding companies. Because of the high price and world recession, market demand has dropped and sales have fallen off. In February 1975, in fact, the lack of both a market and storage space forced OPEC production so low that all its Arab members cut their output substantially *below* the politically inspired cutbacks they had ordered at the height of the 1973 oil embargo. In these circumstances, oil producers are not likely to limit sales further because of dissatisfaction with investment opportunities.

Of course, some countries conceivably could cut output for reasons having little to do with economics, for instance, to preserve their traditional social fabric, or, as in the case of Kuwait in 1972, simply to make their oil reserves last longer. But since other OPEC countries, with more pressing economic needs or more ambitious development programs, are waiting to take up the slack, the world flow of oil should not be affected.* Trouble is more likely to arise from certain aspects of primary recycling that principally affect the financial markets or from secondary recycling.

Primary Recycling: The Special Problems

Each importing nation has an array of assets to offer OPEC but may have trouble in matching them to the investments the OPEC nations would like to make. In overall figures, the industrialized countries would appear to have enough assets to meet the OPEC surplus. Although the cumulative OPEC surplus in 1980 is estimated at $250 billion, the OECD countries produce in excess of an additional $300 billion in marketable financial assets each year. But as of 1974, the OPEC nations had shown a marked preference for liquid assets, choosing these instruments for the bulk of their surplus. Intergovernmental loans and portfolio investments may eventually come to constitute an increasing share of each year's asset accumulation, but at present, loans make up a distant second place to liquid investments, and direct and portfolio investments are a very poor third. This preference for liquid investments has created problems involving the structure of the financial markets, and the consumer coun-

---

Mr. Adelman comments:
    It is counterproductive to aid Saudi Arabia to develop additional producing capacity, since this development will be at the expense of those OPEC countries that have stronger incentives to expand but may refrain from doing so for fear of Saudi competition.

tries' fear of direct investment has raised unwarranted political obstacles.

For a time, bankers worried that they could no longer accept the large volume of short-term deposits from OPEC countries, because their institutions were close to exceeding the maximum prudent ratio of deposits to capital. Banks reinvest their holdings in loans with generally longer maturities and are reluctant to take a high proportion of short-term deposits, because the deposits may be withdrawn before the loans come due. But although the problem of matching maturity schedules may place some strain on the market, it also provides opportunities for profitable adjustment, which the financial markets can be expected to meet with enterprise and dispatch.

If OPEC's financial managers are willing to accept a modest cut in interest for the sake of maintaining a high volume of short-term deposits, financial intermediaries will almost certainly be ready to undertake certain risks of borrowing short and lending long. It would be highly desirable to reduce these risks by designating an official lender of last resort for the Eurocurrency market,* although even in the absence of such a provision, special arrangements can probably be made to meet sudden demands for liquidity. But the most likely outcome is that the financial agencies of the surplus OPEC countries will alter the composition of their holdings as market costs dictate, and that the financial markets will respond, at a price, to the investment preferences of those agencies.

In strictly economic terms, most oil-importing countries should welcome investments from OPEC nations, or anyone else for that matter, as long as the investors do not impose conditions that contravene the national interest or political tradition of the host. OPEC investments help finance the oil deficit. As long as the oil imports are not paid for with exports, the capital outflow from OPEC enables the oil-

*Mr. Flanigan comments:
No single institution is equipped to take on this role. The agreements among central banks mentioned on page 21 of the Report adequately meet this problem.

importing countries as a group to maintain consumption, investment, government purchases of goods and services, and their reserves at the levels they would have occupied if the cost of imported oil had not increased. To the extent that some countries attract OPEC funds in excess of their oil deficits, they will be able to increase their consumption, investment, and government purchases of goods and services through larger imports from other oil-importing countries or alternatively to increase their own foreign investments or reserves. Likewise, to the extent that the oil deficit of other countries exceeds the inflow of funds from OPEC, they will be able to do less than before the price increase.

The Task Force does not see much merit in the widely publicized political objections which have been raised against direct investment. OPEC's few ventures in this line—for example, Iran's purchase of a quarter interest in Krupp's steel-making operations and its proposed but repeatedly postponed $300 million investment in Pan Am—have generated an inordinate degree of public alarm. Further restrictions on foreign investment would be ill-advised. Existing law already protects strategic American industries from foreign control; further measures would have to be framed to affect all foreign investors, not simply those of OPEC, and could lead to retaliation against the far more substantial American investments abroad.

Public attention to these investments, furthermore, has greatly exaggerated their role in recycling. Of the $11 billion that OPEC countries invested in the United States in 1974, only about $100 million, not quite 1 percent, was identified as investment for direct control (defined as equity ownership in real estate or control of more than 10 percent of the voting stock of a corporation).[4] Comparing this figure with the expected volume of trade between the United States and OPEC and its Arab members puts the problem in somewhat clearer perspective.

A more serious problem may arise from the Arab economic boycott against Israel and its secondary application against companies that do business with Israel or have prominent Jewish affiliations. Although this boycott has been vaguely defined and capriciously enforced, its paper provisions call for a broad secondary boycott; companies hoping to trade with the Arab world must not only shun direct investment in Israel but also refuse to deal with other companies on any of the Arab world's boycott "blacklists." Some American businessmen may honor this secondary boycott on their own initiative in the hope of gaining entry to the Arab market. President Ford has already condemned such discrimination,[5] and the government should watch closely for any violations of antidiscrimination or antitrust laws.

The government also may have to strengthen the provisions of the Export Control Act requiring reporting of boycott compliance. This law demands that a company notify the Commerce Department in a confidential report each time it is asked to enforce the boycott; the firm is not required to say whether or not it complied. The law appears to have been widely ignored, but it could effectively deter boycott compliance if the reports, identified by name, were made available to the public.

OPEC financial spokesmen, concerned about adverse public reaction to several cases of direct investment, have complained that they have no clear idea what policy the industrial countries will adopt toward their investments, even though the United States government has already issued clear guidelines for foreign investors. *The Task Force suggests that, in its relations with OPEC investors and indeed with all foreign investors, the United States government should clearly establish the following guiding principle: Foreign investment will be welcome in all but a few clearly designated areas, set aside by law because of their sensitivity, as long as the investor is solely concerned with pursuing legitimate economic ends.* In addition, the American people should be aware that foreign investment does not threaten the national well-being. Foreign-owned businesses in the United States are subject to all United States laws, including antitrust and antidiscrimination laws. Some may have to be strengthened to deal with any political intrusions, for example, attempts to apply the anti-Israel boycott to American business transactions. But such measures need not impair our welcome of OPEC investors.

Safeguarding International Financial Stability

The huge OPEC surpluses themselves might be used as a political weapon, in accordance with either of two scenarios that represent two varieties of a "shifting cargo" problem that might be caused by sudden transfers of OPEC short-term balances. In one scenario, OPEC depositors en masse suddenly demand withdrawal of their deposits, posing a threat to national and international banking systems. In principle, this problem can be solved by a lender of last resort. National banking systems have such a lender. Although the position of Eurocurrency lenders is less clear, the central banks appear to have this matter in hand. In July and September 1974, it was announced that the various central banks would bail out commercial banks that ran into difficulty and that the central banks had apportioned responsibility among themselves for particular commercial banks. If such arrangements prove inadequate, in the very last resort, the crisis could result in a suspension of payments. The OPEC depositors are aware of this possibility and would hardly be likely to take steps that would make it a fact, except possibly in the event of an international political crisis.

The second danger involves massive shifts from one currency to another. Existing swap arrangements could handle substantial shifts for limited periods, through extensions or "turnovers." These arrangements are adequate for normal purposes, the switches of funds associated with speculation (or prudent management) in anticipation of devaluations and revaluations. Some observers fear that OPEC's liquid funds would exceed the amounts that this system can accommodate. But the leading central banks maintain such frequent communications and cooperate so closely that they would certainly be able to handle such a crisis. Under floating exchange rates, furthermore, if massive shifts of liquid funds were not counterbalanced by central bank transfers, they would drive the exchange rate against the party doing the shifting to its cost. In either case, this tactic has little appeal as a political weapon.

PAYING FOR OIL: SECONDARY RECYCLING

For every dollar that OPEC receives in excess of expenditures on exports, some country outside of OPEC must go one dollar in trade deficit. If each oil-importing country received a capital inflow from OPEC equal to its OPEC deficit on current account, it would have no problem in paying for its oil other than the current real burden of its exports and the eventual burdens of debt servicing and repayment. But most countries have in fact been receiving less in capital flows from the OPEC countries than their current-account deficits with those countries. In 1973, $53 billion of the $70 billion OPEC surplus was placed in the Eurocurrency market, the United States, and the United Kingdom, or remained on oil company books as accounts receivable. The reason, of course, is that some countries are less creditworthy, have less well-developed capital markets, or compete less successfully in world trade than others. But such variations are the source of some of the most serious indirect effects of the oil price increase.

A country that is getting less in OPEC capital than it owes OPEC on current account is in primary deficit. (Technically speaking, it is running a negative primary recycling balance on oil account.) In such a situation, the country may follow whichever it can of the following five strategies, some of which involve serious political and economic costs:

1. Primary trade: export more to—or import less from—OPEC countries;
2. Primary capital recycling: procure a larger capital inflow from the OPEC countries;
3. Secondary trade: export more to—or import less from—other non-OPEC countries;
4. Secondary capital recycling: procure a greater inflow from the non-OPEC countries;
5. The final resort: draw down reserves.

Trade strategies that involve restraining domestic demand in order to divert resources from domestic uses to exports may lead to or intensify a recession. Reducing the supply of domestic goods, on the other hand, may intensify inflation. Either result will lead to reduced real income and living standards and possibly to political turbulence. Capital recycling strategies are less immediately painful, but they pile debt on debt and may strain a country's creditworthiness. A country may have to reduce its current deficit through trade simply to maintain its credit standing.

An individual country's choice of strategy will have international as well as domestic repercussions. Changes in the current or capital balances of the primary deficit countries as a group must be matched by equal and opposite changes in the corresponding balances of the surplus countries as a group, OPEC and non-OPEC together. The aggregate current and capital balances of the OPEC countries are largely determined independently of what an individual primary-deficit country does; the effect of anything such a country does (other than curbing its oil imports) must be matched by offsetting effects on the corresponding balance of the other non-OPEC countries.

If a primary-deficit country exports more goods and services and imports less (other than oil), the other non-OPEC countries must export less and import more. If such a country imports less capital, the other non-OPEC countries, jointly, must export less. Some of the most serious problems of secondary recycling derive largely from the unwillingness of the non-OPEC countries to follow consistently complementary strategies.

So far, the worst fears of breakdown in secondary recycling have not materialized. The leading industrial countries, particularly West Germany and the United States, have managed to obtain substantial current-account surpluses as recession diminished their import demands; and capital outflows, stimulated by favorable interest rates in other countries, kept their exchange rates low. Indeed, they have balanced their surpluses by supplying secondary capital recycling. The smaller and less-developed countries experienced this pattern in reverse. In 1974, these countries were able to use secondary capital recycling both to maintain the growth of their imports, in spite of slackened exports, and to avoid trade restrictions and undesired currency depreciation. But now, their debt is piling up, and their current payment balances, which reflect not only the surpluses of OPEC but those of the leading consuming countries as well, cannot long continue for many of them.

Manipulation of currency exchange rates might further destabilize this situation. The current system of floating exchange rates allows the value of each currency to fluctuate against that of others more or less in accord with their comparative international economic position. If these rates floated without impediment, many of the adjustments required in distributing the enormous oil deficit would occur in response to changes in exchange rates, although the fluctuations might disrupt the appropriate pattern of trade and investment. But there is some evidence that exchange rates are being hardened either by direct intervention or by management of domestic demand. In pursuing this course, the leading countries may be merely exporting recession to the smaller countries. The less-developed countries have already had to accept a 10 percent decline in the volume of their imports.*

As the deficit countries fall deeper into debt and experience a decline in imports, some may require intergovernmental help on their secondary capital recycling. The OECD countries are expected to have access to the Kissinger-Simon "safety net," officially called the Financial Security Facility of the OECD, by late 1975. Non-OECD nations can obtain limited but still substantial capital flows on semicommerical terms from the Oil Facility of the International Monetary Fund (IMF). Countries that find these rates still too steep may eventually obtain relief from the "development security facility" proposed for the IMF by Secretary of State Kissinger at

---

Mr. Flanigan comments:
This drop in imports can primarily be explained by economic forces other than currency manipulation, such as reduced export earnings resulting from recession and lower commodity prices.

the Seventh Special Session of the UN General Assembly. The IMF simultaneously moved toward funding this facility when its Monetary Negotiations Committee, meeting over the Labor Day weekend, approved the open-market sale of 25 million ounces of its gold stock; the profits were to be used for the benefit of the developing countries. *The Task Force commends this action and urges the United States government to support expeditious implementation of it.*

Without such aid, some countries may in desperation resort to drawing down currency reserves. This final resort is insufficient to manage a problem of the size and duration of prospective oil deficits and may lead to the destructive sequence of devaluation and competitive export drives, trade restrictions, and limitation of domestic demand.

Somewhat surprisingly, the primary deficit countries have not yet drawn much on their reserves to meet the oil price rise. They resorted to borrowing instead. *The Task Force believes that, even in the face of poor creditworthiness, secondary capital recycling through official agencies would be justified for the sake of avoiding the vicious cycle described above.* Of course, countries whose current-account imbalance reflects more than their oil deficits should not depend on such recycling; they should be taking painful but ultimately necessary steps to remedy the situation.[6]

Faced with excessive debt or limits on credit, primary deficit countries may seek relief through trade recycling. If many non-OPEC countries choose this course, they may find themselves engaged in a senseless tug-of-war, unless the relative surplus countries—those with actual surpluses or with deficits that are small relative to their creditworthiness—follow a complementary policy of allowing their own current balances to be reduced, even into deficit, which, after all, they are in a better position to finance. Through such an accommodation, the surplus countries could allow the deficit countries to improve their trade balances to the point where their credit would be reestablished, and they could finance the remainder of their deficits by capital recycling. Secondary trade recycling is particularly desirable in those cases where the resulting trade patterns are likely to be sustainable beyond the period of OPEC imbalance. Economic recovery may partly restore the balance between non-OPEC surplus and deficit countries by increasing the imports of the former. But it is important that no impediments be placed in the way of trade policies needed to complete the job.

The fundamental problem remains that someone must bear the deficits that are the counterpart of the OPEC surplus. Under these circumstances, policies that normally do not have beggar-thy-neighbor consequences may acquire them. Failure to recognize this problem, and such recognition is not evident in current policies, will adversely affect international trade. *The Task Force urges therefore that, as secondary recycling proceeds, all non-OPEC countries, surplus and deficit alike, assume the responsibility of working toward mutually consistent goals, with free movement of goods and services and capital in response to market conditions.*

THE FOURTH WORLD[7]

The oil price rise has brought fabulous wealth to a few Third World countries, the so-called rich-poor nations, but it has consigned those Third World countries without oil to a new category in world affairs—the Fourth World, or poor-poor nations. This category must in turn be divided into two groups: the not-so-poor and the poorest-of-the-poor. If the dividing line is a 1972 per capita income of $200, the poorest group contains 23 countries, with an average per capita income of $110, and the not-so-poor contains 71, with an average around $530. The poorest countries have an aggregate population of about 900 million, 35 percent of the population of the entire noncommunist world, but they account for only 3 1/4 percent of its aggregate Gross National Product (GNP). Most of these countries are located in South Asia and East and Central Africa. The three countries of the Indian subcontinent and nearby Sri Lanka make up 81 percent of their population.

As a group, the Fourth World weathered the initial impact of the oil price increase quite well.[8] Although their oil imports were estimated to have cost an additional $10 billion in 1974, they were able not only to absorb the blow but to increase their volume of imports as well. The source of this temporary good fortune was the unusually high prices that Fourth World exports commanded in 1973, which actually allowed them to accumulate reserves. These countries made it through 1974 by drawing down reserves and increasing the nominal amount of their aid receipts and borrowing by 50 percent. But these aggregate figures conceal some deplorable situations which require emergency aid.

Furthermore, the financial situation of the Fourth World grew worse in 1975. Recession in the industrial world has reduced the market for Fourth World exports, and inflation has increased the cost of their imports. In the absence of a substantial inflow of capital, not yet in evidence, the Fourth World will have to reduce its imports. Even with a strong OECD growth rate after 1975, the real Fourth World import capacity is expected to grow at only 7 percent a year. Under these somewhat optimistic assumptions, between 1976 and 1980, the Fourth World per capita GNP is likely to grow at a rate of only 2.1 percent a year, a substantial drop from 4 percent in 1973 and an average of 3.2 percent between 1969 and 1973. Moreover, although the not-so-poor group is expected to grow at 2.5 percent, the poorest group will grow at less than 1 percent. At that rate, their per capita incomes will take three generations to double.

This deceleration of growth will result not from the oil price increase and the decline in basic commodity prices but from the expected failure of real capital flows to grow in proportion to Fourth World economies. The shortfall is expected to total $20 billion a year (in 1974 dollars) over the next five years. If this sum were forthcoming and other conditions were favorable, the Fourth World could grow at a rate of 3.3 percent, which would be consistent with the growth target of the second development decade. The poorest countries would require only $2 billion more (in 1974 dollars) to support a program to raise their growth rate from 0.9 percent to 2.0 percent.

The most important single thing the OECD countries can do for the Fourth World is to continue to prosper. The purpose of this apparently selfish advice is to guarantee flourishing markets for Fourth World exports. In fact, OECD–Fourth World relations provide a textbook example of the gain from international trade. Specifically, the OECD should grant freer entry to raw materials and simple manufactures from the Fourth World. These earnings will increase the demand for goods from the developed countries.

In particular, the developed world ought to lower the barriers against those simple manufactures that the developing world is beginning to export. These industries are at a comparative disadvantage in the economies of the developed world; the countries concerned should phase them out and adopt measures to assist the transition for workers and resources.

Concern for the developing nations has been expressed from the outset of the current "Tokyo Round" of negotiations among the signatories of the General Agreement on Tariffs and Trade (GATT). The participants have agreed to maintain and improve the Generalized System of Preferences. For this purpose, the developed countries will have to raise the quantitative limits imposed on some imports from Fourth World countries that receive preferential rates. Furthermore, the general reduction of duties may erode the value of these Generalized Preferences. But the Fourth World is likely to gain on balance from the results of the Tokyo Round. They may benefit most from the elimination of certain non-tariff barriers, especially import quotas.

*Trade liberalization is superior on all counts to commodity agreements, which are likely to be short-lived, costly to at least one side, and encouraging to the misuse of resources.** As Secretary Kissinger wisely noted in his speech for the Seventh Special Session of the UN General Assembly, the aim of American economic policy should be "to safeguard and improve the

Mr. Nathaniel Samuels comments:
Despite the deficiencies of commodity agreements, we should not be precluded from continuing to explore practical means for stabilizing the supply of basic commodities.

open trading system on which the future well-being of all our countries depends."

But the Fourth World may have trouble managing its resources to meet the enlarged export market. And the temptation will be strong to use the proceeds from export sales to pay for increased imports for consumption rather than development. Although the Fourth World would certainly benefit from higher levels of consumption, this possibility means it will be harder to finance growth through exports than through equal flows of capital.

The OECD countries are more likely than OPEC to try to increase capital flows above present projections. In fact, the official OPEC contribution is highly unlikely to exceed the $10 billion assumed by the World Bank. About $2.5 billion of this will consist of bilateral aid, principally between politically affiliated countries like Saudi Arabia and Egypt. Another $5 billion to $6 billion should flow through multilateral agencies like the IMF oil facility, although the full OPEC commitment for 1975 is yet to be announced, and these loans are made at nearly commercial rates. If the annual OPEC surplus continues to shrink, special efforts will be required to keep its contribution at this level. Already, there are some signs of tightening up on commitments.

Furthermore, inflation is likely to neutralize the increase in official capital flows from the industrial world. It seems unrealistic to hope that either OPEC or the industrial world, or both together, will adopt a program of aid to the Fourth World that would restore the target of 6-percent growth in aggregate output set for the second development decade. But more can be done than now seems likely. Although an increased capital flow could not automatically increase the growth rate, it is necessary for any program that could. (On the other hand, the Fourth World countries will have to organize themselves to put any aid to effective use. Without this effort, future aid may simply be the palliative it has often been in the past.)

The Task Force commends Secretary Kissinger's attention in his Seventh Session speech to the question of providing the additional financing needed for development of the poorest countries. We applaud the recent progress toward the IMF trust fund. We also suggest that the industrial countries, OPEC, and the Fourth World should explore a triangular arrangement for development. OPEC would supply capital, sharing the risk of default with OECD countries. Subsidized interest would return individual OPEC investors a commercial or near-commercial rate. The OPEC and the OECD countries would share the cost of the subsidy, and the Fourth World borrowers would receive highly concessional terms. The Fourth World would spend these loans in OECD countries, which would then have additional means to finance their deficit with OPEC.

The Task Force strongly recommends enlarged aid to the poorest Fourth World countries, whether by this plan or another. The need is great, and OPEC has indicated an interest. Such measures would relieve the immediate distress of the poorest lands, renew hope for eventual improvement of their living standard, and provide a constructive meeting ground for the OPEC and OECD countries. The proposed arrangements seem to offer gains to all parties within a benevolent framework.

ADDITIONAL VIEWS

Walter J. Levy comments:

The Task Force Report gives insufficient attention to two points. The first is the significance of differences among the members of OPEC. Although a few OPEC countries, particularly Saudi Arabia, will continue to accumulate substantial current-account surpluses over the next few years, many other OPEC countries will soon experience current-account deficits. This difference may lead the OPEC nations to pursue different interests and perhaps different policies not only with regard to oil-production levels and prices but also with regard to general financial matters and overall development programs. The oil-importing countries may then find themselves dealing with the continued financial accumulations of some OPEC countries, while helping other OPEC countries to finance imports. The economic, political, and strategic military relevance of these developments certainly warrants further analysis.

The second point that the Report does not cover is that even when aggregate OPEC imports equal or exceed OPEC current revenues (the different financial positions of OPEC countries aside), the OPEC countries will be obtaining their imports from a relatively small number of mostly OECD countries. Even though the benefits of such increased trade between a few OECD and OPEC countries may trickle down to other OECD and developing countries, the size and direction of the trickle will not correspond in either timing or amount to the oil-trade deficit of the respective countries. Accordingly, secondary recycling, as the Report describes it with reference to the period when OPEC is accumulating financial surpluses, will still be necessary when OPEC needs all its current revenues for imports of goods and services.

Mr. Adelman and Mr. Nathaniel Samuels wish to associate themselves with this statement.

NOTES TO TASK FORCE REPORT

[1]See Background Paper, Table 12, p. 79.

[2]OECD, *Economic Outlook*, no. 17, July 1975.

[3]Yusif Sayigh, quoted in Robert Mabro, "Can OPEC Hold the Line," *Middle East Economic Survey* (supplement), February 28, 1975, and an anonymous OPEC official quoted in *Petroleum Intelligence Weekly*, March 10, 1975.

[4]Sanford Rose, "The Misguided Furor about Investment from Abroad," *Fortune*, May 1975, p. 172. This figure does not count Saudi Arabia's takeover of a large portion of Aramco as direct investment in the United States.

[5]*New York Times* , February 27, 1975, p. 1.

[6]Ceteris Paribus, pseud., "Vicious Recycling," *Foreign Policy*, no. 17, winter 1974–75, pp. 85–87. The author is reported to be a "high treasury official."

[7]The projections in this section are based on the World Bank staff reports of July 8, 1974, and April 8, 1975. They indicate the nature and magnitude of the problem under assumptions detailed in the source documents. The projected figures should be taken not as unconditional forecasts but as illustrations of possible future trends.

[8]This observation refers only to international financial developments. Some of these countries also had disastrous crop years.

Background Paper

By Sidney S. Alexander

I/The Energy Crisis and the Cartel Problem

No world oil shortage exists in the short run nor energy shortage in the long run. As of the fall of 1975, there is a worldwide surplus of oil-producing capacity, exceeding present output, estimated at 11 to 12 m.b.d. compared with current OPEC exports of less than 26 m.b.d. This surplus capacity could easily be enlarged by further development of known reserves. Furthermore, new oil reserves are being proved faster than the old are being tapped. Clearly, physical availability alone is not the problem.

The problem is the tight control exercised over the international flow of oil by the 13 members of OPEC, particularly the Arab oil producers, who control 63 percent of OPEC production capacity and are committed to the use of oil as a political weapon in the struggle with Israel. The cartel's success hinges on two facts: inelastic demand for crude oil, and the inability of the oil-importing countries to increase their domestic production of oil and oil substitutes.

The conditions underlying inelasticity of demand are also responsible for the power of oil as a political weapon. A cutoff of Arab oil would cripple the economies of Western Europe and Japan, whose domestic oil supplies and alternative sources are far short of their requirements. In practical terms, therefore, a solution to the problem depends either on the safeguarding of imports or on the development of domestic supplies of oil and alternative energy sources.

This study focuses on the problems created by the oil-producing countries' temporary inability to dispose of the income generated by their exports and on the oil-importing countries' difficulties in paying for their imports.

PRICE

The oil-importing countries are paying over $100 billion more per year (calculated in 1975 dollars) to the OPEC countries at current prices than under the rates current in 1970 (see Table 6). It is, naturally, important to them to try to lighten this burden. Some observers, including the United States government policy makers, feel that the price of oil can be curbed, and possibly reduced, by negotiation with OPEC. Others believe that the price can be reduced only by weakening OPEC, principally through expansion of non-OPEC energy supplies but also by demand reductions. Until the control of OPEC is weakened, they maintain, we must do our best to cope with the high price.

FINANCING OPEC SURPLUSES

The problem of financing OPEC surpluses begins with the inability, presumably temporary, of the most important oil-producing countries to accept imports equal in aggregate value to their swollen receipts from exports. Until imports come to balance exports (they seem unlikely to do so on net balance until the 1980s), financial instruments and practices are required that will permit the oil-importing countries to finance a substantial part of their imports on credit or through the transfer of assets. The exporting countries must take their excess receipts in the form of claims on the rest of the world—as foreign currency holdings or other reserves or as loans or investments. This process of capital, or monetary flows, the counterpart of the current balance surplus of the oil-exporting countries, is called "primary recycling."

It has been claimed that if the oil-exporting countries do not find attractive monetary, credit, or investment instruments in which to place their excess earnings, they may cut back their oil exports. New arrangements may be needed to accommodate a volume of outstanding claims of oil exporters against oil importers unprecedented in the history of international finance, so that the oil may continue to flow. In addition to the fundamental primary-recycling problem, the provision of sufficiently attractive investment outlets to absorb the OPEC surpluses without impairing the flow of oil, primary recycling presents a number of special problems. These concern the difficulties of matching the supply of monetary, credit, and capital instruments to the OPEC countries with the demands of the latter for such instruments.

Table 1   Production and Installed Capacity of OPEC Countries, 1974 and 1975

	PRODUCTION, 1974 AVE., M.B.D. (1)	PRODUCTION, 1975 FIRST HALF, AVE., M.B.D. (2)	PRODUCTION, PRE EMBARGO,† M.B.D. (3)	INSTALLED CAPACITY, EARLY 1975, M.B.D (4)	EXCESS CAPACITY, 1975: I (PERCENT)‡ (5)
Saudi Arabia*	8.50	7.10	8.60	10.50	32
Iran	6.00	5.50	5.80	6.50	15
Kuwait*	2.50	2.20	3.50	3.80	42
Venezuela	3.00	2.70	3.40	3.30	18
Libya	1.50	1.10	2.30	3.00	63
Iraq	1.90	2.20	2.10	2.60	15
Nigeria	2.30	1.70	2.10	2.50	32
United Arab Emirates	1.70	1.40	1.70	2.20	36
Indonesia	1.40	1.20	1.30	1.50	20
Algeria	1.00	0.90	1.10	1.10	18
Qatar	0.52	0.44	0.61	0.65	32
Ecuador	0.18	0.16	0.24	0.26	38
Gabon	0.18	0.22	0.16	0.19	−δ
OPEC Total	30.60	26.80	32.90	38.10	30

*Including half of Neutral Zone
†September 1973
‡As a percentage of capacity
δProducing above earlier capacity of early 1975

Sources: "1974 Production and Early 1975 Installed Capacity," *Petroleum Intelligence Weekly,* April 21, 1975. "1975: I Production," *Petroleum Economist,* September 1975.

PAYING FOR OIL

Even if the primary-recycling problems were completely solved, many countries might still encounter serious difficulties in paying for oil. If the oil-exporting countries balanced their primary recycling among the oil-importing countries in strict proportion to the oil deficit of each, these difficulties would not arise. If, however, primary recycling is unbalanced in favor of the leading countries, or the Eurocurrency market, then ways must be found for the countries who receive less primary recycling than they require to acquire the needed funds either through capital flows or through trade. This process is called "secondary recycling," or, less sympathetically, reshuffling. The bulk of secondary recycling will be required by developed countries of limited creditworthiness. Their credit problems spring from limitations of liquidity rather than from fundamental insolvency. For them, the problem is to devise ways to safeguard foreign loans and investments in their financial and capital markets from default due to illiquidity. For most of them, too, there is the problem of reducing their current deficits on non-oil account, which is principally a problem of the level of their exchange rate and the management of their domestic demand. Trade restrictions in other countries may intensify this problem.

For still other countries, the poorest of the Fourth World, the problem is to negotiate loans and grants on concessionary terms, for normal commercial terms are beyond their means. The elimination, or favorable modification, of trade restrictions would also help this group.

If the secondary-recycling problem is not solved, many oil-importing countries may exhaust their available credit and be forced to restrict not only oil imports but all other imports as well. One country after another may adopt beggar-thy-neighbor policies, generating a spiral of foreign trade restrictions. Even without explicit trade restrictions, *any* action by one non-OPEC country to reduce its deficit on current account will be at the expense of the other non-OPEC countries as a group, with a given OPEC surplus on current account. Some countries already have tried to improve their balance of payments by restricting demand, inspired, in large part, by the inflationary trends which were already full-blown before the great oil price rise of the winter of 1973–74. Those restrictive demand arrangements have brought on the deepest worldwide recession of the postwar period.

The most serious consequence of the oil price rise has been its tendency to deepen and prolong this recession. The price rise itself has intensified both the rate of inflation and the counterinflationary restrictions on demand. By diverting consumer expenditures from other products to oil, the price rise reduced demand for other products. It also reduced the demand for energy-using equipment, primarily automobiles. This was only partially counterbalanced, at least over the short term, by the stimulation of investment in the energy field. Such investment, while potentially significant, has been slow in developing.

II/*The Problem of Price*

World crude oil production in 1973 and 1974 ran at 58 m.b.d., of which close to 31 m.b.d. were from OPEC countries (see Table 2). Most of the balance was consumed in the country of origin, so that almost all oil in international trade came from the OPEC countries. The OPEC countries account for two-thirds of the entire world's petroleum reserves, 80 percent of the noncommunist world's reserves (see Table 2). Furthermore, the prospects of finding and proving additional reserves in the OPEC countries, particularly in the Persian Gulf, are more promising than elsewhere, and the cost of exploration per barrel discovered is miniscule. The oil, when found, is cheaper to produce—about $0.25 to $0.50 a marginal barrel in the Middle East compared with about $4 a marginal barrel in the North Sea and $12 in the United States. The long-run cost of transportation from the Persian Gulf to the United States or to Western Europe is approximately $1 (in 1975 dollars) per barrel. Thus, a commodity that can be produced and transported for less than $1.50 a barrel is selling for about $12.50 delivered.

Saudi Arabia is so large a supplier, both actual and potential, that if it holds marker crude (light Arabian 34°) oil at its late 1975 price of $11.51 a barrel (1975 dollars) at the Persian Gulf, the real price of oil can be held constant. Any movement of the price of marker oil, on the other hand, can be regarded as a shift in the *level* of world oil prices. OPEC price resolutions fix the price of marker crude in dollars and cents. Other crude differentials are covered only by statements of general

Table 2   World Oil Tableau, 1974

	PRODUC-TION (M.B.D.) (1)	CONSUMP-TION (M.B.D.) (2)	NET EXPORTS(+) OR NET IMPORTS(−) (M.B.D.) (3)	PUBLISHED PROVED RESERVES (BILLIONS OF BARRELS FOR END 1974) (4)
United States	10.5[a]	16.2	−5.9	40.6
Canada	2.0	1.9	+0.2	8.8
Western Europe	0.4	14.2	−14.4	26.3
Japan	—	5.3	−5.4	—
Australasia	0.4	0.7	−0.3	2.4
OECD	13.3	38.3	−25.8	78.1
Other Developed Countries	—	1.0	−1.0	—
OPEC Countries	30.7	1.1	+29.5	484.9
Saudi Arabia[b]	8.5	0.07	+8.5	173.1
Iran	6.0	0.30	+5.7	66.0
Venezuela	3.0	0.25	+2.7	15.0
Kuwait[a]	2.5	0.02	+2.5	81.4
Nigeria	2.3	0.05	+2.2	20.9
Iraq	2.0	0.10	+1.7	35.0
Libya	1.5	0.04	+1.5	26.6
United Arab Emirates[c]	1.7	0.01	+1.6	33.9
Indonesia	1.4	0.19	+1.2	15.0
Algeria	1.0	0.06	+0.9	7.7
Qatar	0.5	—	+0.5	6.0
Gabon	0.2	0.01	+0.2	1.8
Ecuador	0.2	0.03	+0.2	2.5
Non-OPEC Middle East Exporters[d]	0.4	0.01	+0.4	6.3
Less-Developed Country Importers	3.4	6.1	−2.8	39.7
World (excl. comm. countries)	47.5	46.5	+30.1[e] −31.1	609.0
Communist countries	10.5	9.5	+1.0	111.4
USSR	9.1	6.9 }	+1.0	83.4
Eastern Europe	0.4	1.6 }		3.0
China	1.1[f]	1.0[g]	—	25.0
World	58.0	56.0	±31.1[h]	720.4

[a]Including natural gas liquids (1.7 m.b.d.)
[b]Including share of Neutral Zone

principle to the effect that they should match the price of marker crude with due allowance for quality and location. Consequently, in practice, the differentials are set by unilateral decisions of the OPEC members concerned, subject only to informal retroactive influence by the other members. Changes in the prices of other crudes relative to the price of marker crude can, accordingly, best be regarded as adjustments of the price *structure* but not of the price level. The other OPEC countries have scaled their prices to parallel the Saudi Arabian price for marker crude, taking account of transportation and quality differences. Occasionally, a member nation makes a price adjustment, but by and large these adjustments are the exception that prove the rule.

For example, Libya found in March 1975 that its oil, priced at a differential from marker crude based on higher transport rates, was out of line after tanker rates fell to a fraction of their previous level. It gradually and reluctantly moved its price downward about 30 cents a barrel. This was widely hailed as a sign of a break in the price level. It was, of course, no such thing, but merely an adjustment in the price structure relative to an unchanged level of the price of marker crude. Output, which had fallen below 1 m.b.d. in March, recovered to 1.5 m.b.d. in June. Capacity is put at 2.4 to 3.0 m.b.d.[1]

Similarly, Abu Dhabi found in February 1975 that its sales, which had increased 1.7 m.b.d. in 1974, had fallen to 0.75 m.b.d. It reduced quality premiums for its oil by $0.25 to $0.60 a barrel, and output rapidly recovered.

The oil cartel of the OPEC countries functions, in effect, through adherence to a semiofficial design of government per-barrel levies, or "takes." These takes, together with small adjustments in the companies' margins, dictate the prices of

---

[c]Abu Dhabi, Dubai, and Sharjah are the producers in the United Arab Emirates

[d]Principally Oman and Bahrein (producing 0.25 and 0.068 m.b.d., respectively, in 1974)

[e]Total net imports or exports

[f]Includes Albania

[g]Includes Albania, North Korea, and North Vietnam

[h]Difference between production and consumption is accounted for by stock changes and unknown military liftings.

the various crudes. The pricing scheme, together with the pattern of the oil companies' distribution systems, determines the sales volume of the respective producing countries. The Saudi Arabian take on marker crude was pegged, as of October 1, 1975, at $11.17 per barrel by the Saudis. Add a production per-barrel cost of $0.12 plus a company margin of $0.22 per barrel, and the result is the current $11.51 price of light Arabian 34° at Ras Tanura on the Persian Gulf.

The current organization of the world oil market can accordingly be described as a cartel composed of OPEC members, the governments of the 13 principal oil-exporting countries. Output is not explicitly allocated by the cartel, but, essentially, it is governed by scaling the prices of all other crudes to the current price of Saudi Arabian marker crude. Relative output fluctuations do occur, largely governed by changes in market conditions, such as transport costs, that render inappropriate differentials that reflect earlier conditions.

PRICE BEHAVIOR BEFORE 1973

This was not the structure of the world oil market prior to October 1973. The principal formal difference is that before that date the government takes were negotiated with the companies instead of being set unilaterally by the governments of the producing countries. The principal practical difference is that the price setting was much more a company affair, against the background of the cost of oil to the company of currently negotiated take plus production costs, which were, over that period, only a few cents a barrel, and the market conditions set by the actions of other companies.

In Saudi Arabia, the government take was set in 1933 at four gold shillings per ton, which over most of the period between 1943 and 1949 worked out at $0.17 to $0.24 a barrel (see Table 3). Then, about 1950, a new formula for the government take was widely adopted: a 50-50 share of the profit. Thus with Persian Gulf oil selling at about $1.80 a barrel and with costs at about $0.12 a barrel, this implied a take of about $0.83 a barrel for the Persian Gulf countries (see Table 3).

The 50-50 formula succeeded, almost miraculously, in giving the producing countries a substantial increase in take without cost to the consumers or to the companies, the consequence of a favorable interpretation of United States tax law, soon copied by Britain. Tax paid to the producing countries by the oil company could be credited against its home-country tax liability. By labeling most of the take as tax, the increased cost of the 50-50 split was translated into a reduction of the tax payments to the offtakers' home countries— the United States and Britain.

From the end of World War II until the end of 1970, the price of oil in the Middle East trended downward from about $2.22 in early 1948 to about $1.25 in 1970 (see Table 3). Occasional upswings, as, for example, to $2.08 in 1957 after the Suez crisis and to $2.54 immediately after the 1967 war, were but temporary plateaus in the price erosion. In real terms, of course, the price erosion was even greater. The precise course of prices is difficult to document because the prices actually paid and received are not matters of public record. "Posted prices" are public, but it must be emphasized that the posted price is not really a price. It is a reference number for certain calculations, particularly for the determination of the government's take. It has also been used as a "transfer price" for accounting and tax purposes by the major affiliates of the producing companies. The closest thing to a true market price for Mideast oil over much of the period from 1960 to October 1973 is a computed netback price derived from the product prices in the Rotterdam market.[2]

In the early postwar period, the oil companies often tied the price of Middle East crude to United States prices, first by simply charging the same price at the Persian Gulf as at the Gulf of Mexico and then by applying more sophisticated netback formulas. Pressure from the U.S. Marshall Plan Agency (the 1948 predecessor of AID), which was paying for substantial imports into Europe, led to a reduction from over $2 in 1948 to about $1.65 in late 1949.[3] Thereafter, prices trended

Table 3   Prices and Government Takes: Oil at the Persian Gulf[a]

DATE	ESTIMATED MARKET PRICE (CURRENT $ PER BARREL) (1)[b]	ESTIMATED MARKET PRICE (1975 $ PER BARREL) (2)	POSTED PRICE (CURRENT $ PER BARREL) (3)	TAKE–SAUDI ARABIA (CURRENT $ PER BARREL) (4)	TAKE—SAUDI ARABIA (1975 $ PER BARREL (5)
1945	1.05	3.28	—	0.17	0.53
Dec. 1947	2.22	5.53	—	0.17	0.42
Mid- 1948	2.08	4.87	—	—	—
Mid-1949	1.85	4.35	—	—	—
End-1949	1.65	3.86	—	—	—
End-1950	—	—	1.71	0.28	0.63
End-1953	—	—	1.93	—	—
June 1957	2.08[c]	3.92	2.08	0.80	1.51
End-1959	—	—	1.90	0.74	1.34
End-1960	(1.50)	2.68	1.80	0.69	1.23
Avg. 1961	(1.50)	2.67	1.80	0.70	1.24
Avg. 1962	(1.61)	2.83	1.80	0.69	1.21
Avg. 1963	(1.59)	2.76	1.80	0.70	1.22
Avg. 1964	(1.29)	2.21	1.80	0.77	1.32
Avg. 1965	(1.17)	1.96	1.80	0.78	1.31
Avg. 1966	(1.27)	2.07	1.80	0.83	1.36
Early 1967	(1.28–1.23)	(2.04–1.96)	1.80 ⎫	0.88	1.39
Late 1967	(0.96–2.54)	(1.50–3.98)	1.80 ⎬		
Avg. 1968	(0.90–1.83)	(1.37–2.78)	1.80	0.84	1.28
Avg. 1969	(0.71–1.27)	(1.03–1.84)	1.80	0.82	1.19
Early 1970	(0.51–1.44)	(0.71–2.02)	1.80 ⎫	0.89	1.25
Late 1970	(0.14–2.01)	(0.19–2.71)	1.80 ⎬		
Feb. 15, 1971	1.65	2.20	2.18	1.19	1.58
Jan. 20, 1972	1.85	2.38	2.48	1.45	1.87
Jan. 1, 1973	2.20	2.73	2.59	1.52	1.89
April 1, 1973	2.30	2.80	2.76	1.63	1.99
June 1, 1973	2.70	3.26	2.90	1.71	2.06
Oct. 16, 1973	3.65	4.32	5.12	3.05	3.61
Jan. 1, 1974	8.45	9.75	11.65	7.00–9.31[d]	(8.08–10.74)
May 1974	9.55	10.62	11.65	9.31[e]	10.36
Dec. 1974	10.46	10.84	11.25	10.12[e]	10.48
Oct. 1975	11.51	11.30	— [f]	11.17[e]	10.96

[a]Based on Arabian light 34° at Ras Tanura.

[b]Figures in parentheses represent Rotterdam product prices netted back to the Persian Gulf. When two figures are given, the first represents the use of a spot tanker rate, the second, a long-term tanker rate. After the 1967 war, when tankers were in great demand, the low netback reflected a penalty for not having tankers on a long-term

slowly upward, sharply accelerating during the Suez incident of 1956–57.

By the conclusion of the Suez affair, the market price was over $2; a sharp decline followed until 1960, and thereafter a gradual erosion prevailed through the sixties, interrupted, but not long reversed, by the 1967 war (see Table 3). The posted price for Arabian light at the Gulf was reduced from $2.08 in June 1957 to $1.80 in September 1960. There was probably no discounting at the earlier date and substantial discounting at the later date, so the market price must have declined even more than did the posted price. The Rotterdam-based netback price for Persian Gulf crude in 1960 was about $1.45 or $1.50. We may, accordingly, infer a decline of the market price by over 25 percent from late 1957

---

lease, the high netback reflects an "economic rent" on having a long-term lease. Each number represents the market price at the Gulf for the respective buyers. See M. A. Adelman, *The World Petroleum Market* (Baltimore: The Johns Hopkins University Press, 1972), pp. 190–191.

[c]Author's estimate based on inference that when posted price was moved up to $2.08 it represented a market price.

[d]The $7.00 represents the take on Aramco equity crude only; on government crude, the take was 93 percent of posted price. The $9.31 represents the average take on a 40-60 basis.

[e]The average take computed as in footnote *d*. IMF Survey, Oct. 13, 1975, p. 305.

[f]At the September 1975 meeting, OPEC members seem to have dropped the use of posted prices completely. Posted prices were not mentioned in the resolutions and communiques. *Petroleum Information Weekly*, Oct. 6, 1975.

Sources: Column (1): for 1945, Helmut J. Frank, *Crude Oil Prices in the Middle East* New York: Praeger, 1966), p. 29; for 1947–49, Adelman, p. 134; for 1957, author's estimate; for 1960–66, Adelman, p. 183; for 1967–1970, Adelman, p. 190; for 1971 to May 1974, World Bank Staff, as reported in Joseph A. Yager and Eleanor B. Steinberg in *Energy and U.S. Foreign Policy* (Cambridge, Mass.: Ballinger, 1974), p. 237; for Dec. 1974 and Oct. 1975, current news reports.

Column (2): calculated from column 1 data through use of Department of Commerce implicit price deflator for GNP 1975, with the deflator taken at 186.1 on the basis of 1958 = 100.

Column (3): primary source is *Platt's Oil Price Handbook, 1973*; later sources are from trade press.

Column (4): for 1945, reported to be the same as 1947; for 1947–70, Adelman, p. 208; for 1971, April 1, 1973, and June 1, 1973. estimated by author from contract terms; for Jan. 20, 1972, Jan. 1, 1973, and Oct. 16, 1973, Petroleum Press Service, Nov. 1973; later figures from trade press.

Column (5): calculated from column 4 in the same manner as column 2 was calculated from column 1.

to 1960. In real terms, adjusted for the general price level, the decline was 32 percent.

It was the reduction of the posted price from 1957 to 1960 that encouraged the birth of OPEC. That organization succeeded in stabilizing the posted price at $1.80 through the sixties. Meanwhile, the market price continued, with some fluctuation, to erode in current dollars—and even more in terms of constant dollars. In real terms, therefore, the market price of Persian Gulf oil in early 1967 was only 42 percent of the 1948 price. The consequence was a double squeeze on the oil companies. The government take in Saudi Arabia rose $0.19 a barrel from 1960 to 1967, while the offtakers' net-back, before the June war, had fallen from about $1.50 to about $1.23. The rate of profit on net assets of United States direct investment abroad in petroleum fell from well over 20 percent in the 1950s to 11 or 12 percent in the late 1960s. The latter figure was not noticeably higher than the return on manufacturing at home or abroad.[4]

In late 1970, the first steps were taken that led to the great price rise of 1973–74. The stage was set by the closing of the Suez Canal in the 1967 war and a consequent tanker shortage brought on by the elongated haul around the Cape of Good Hope. Then in May 1970, Tapline, normally capable of bringing half a million barrels a day of Saudi oil to the Mediterranean, was cut by a bulldozer accident in Syria, and the Syrian government claimed that conditions were too turbulent near the border, where the break had occurred, for a repair team to work. Some observers connect this solicitude with Syria's concurrent negotiations for rescheduled transit fees for the pipeline. Nigeria was torn by civil war. The new revolutionary Libyan government, claiming to be displeased with the conservation practices of the oil producers, ordered production cutbacks totaling approximately 400,000 barrels daily. The Libyans concurrently demanded improved terms from the oil companies. With tanker rates sky-high, Libyan oil was selling at a substantial premium, and the Libyans were demanding a modest share of that premium. In September 1970, the Libyan government won a $0.30 increase in the posted price and an increase in the tax rate from 50 to between 54 and 58 percent.

In December 1970, Venezuela unilaterally increased its tax rate from 52 to 60 percent. The Persian Gulf producers, witnessing the Venezuelan and Libyan triumphs, demanded more. Their demands were answered by a series of counteroffers, but at the OPEC meeting at Caracas in December 1970, OPEC issued Resolution XXI-120 calling for, among other things, a "minimum" 55 percent tax rate for all countries, the elimination of "posted price disparities" by tieing the other crudes to the highest existing posting, "with appropriate allowances for quality and location, and a 'uniform price increase' to reflect general market improvement."[6] That this meant, in the first instance, the passing on of Libya's gains to the Gulf producers was clear from another provision of the resolution that called for negotiations between the Persian Gulf nations and their producing companies in Teheran within 31 days. A subsequent OPEC meeting was scheduled within about three weeks of the initiation of negotiations to evaluate the results.

After an extensive and stormy process of negotiations, an agreement was reached in February 1971 at Teheran between the Persian Gulf producing countries and their companies, raising the posted price by $0.35, with an additional $0.11 annual escalation through 1975. The tax rate was increased from 50 to 55 percent. The Saudi Arabian take, for example, was raised from $0.88 to $1.26 per barrel.

The Libyans then announced that if the Persian Gulf countries got more, they, the Libyans, should receive still more. The Persian Gulf states agreed, however, not to leapfrog. So the episode ended with the Tripoli agreement of April 2, 1971, in which the Libyans received an additional $0.90 on the posted price, while standardizing the various tax rates into a 55 percent rate, uniform with the Gulf rates. The per-barrel take was raised from about $1 to $1.79 by the two Libyan agreements. The tax-paid cost of Libyan crude thus ended up $0.80 to $0.90 higher than the tax-paid cost of a comparable Gulf crude.[7]

Venezuela then promptly legislated additional tax and posted price increases; its take per barrel was increased by the sequence of changes from $1.04 in 1969 to $1.72 in 1972.[8]

The oil companies wanted to preclude leapfrogging by jointly negotiating a single contractual rate with OPEC. The United States government cooperated at this point by waiving antitrust objections, but at a critical juncture it saw fit to inform the key OPEC countries that it did not necessarily favor a uniform agreement. Some critics accuse the government of undermining the companies' efforts.[9]

But even if the leapfrogging had been halted, the fundamental change that underlay the oil revolution could not have been avoided. For the first time, the OPEC countries, particularly the Persian Gulf producers, demonstrated their readiness to restrict output. The resulting bargaining strength, rather than any supply shortage in 1970, explains the 1971 and subsequent OPEC successes. Thus, in 1975 the appearance of a 12 m.b.d. overcapacity, about twice as large as the overcapacity of the late 1950s and early 1960s that led to the softening of the market up to the 1967 war,[10] failed to weaken the producing countries' resolve to stand together. In fact, some minor downward adjustments in take were made by those countries—Libya, Abu Dhabi, Nigeria, Indonesia, and Ecuador—whose sales suffered the most, relative to capacity. But the price level, reflected in Saudi marker crude, stood firm. Indeed it was raised in September 1975. Some analysts estimated that the OPEC countries could have withstood a surplus capacity of 16 to 18 m.b.d. before running into the "danger zone."[11]

The increases of producer-country revenues from the Teheran-Tripoli agreements of 1971 were substantial. One contemporary observer estimated that the 1975 revenues of OPEC members would thereby be increased from $10 billion to about $18.5 billion.[12] In fact, of course, the 1975 revenues will be about $100 billion. But, as that observer noted, the principal effect was the changed company-country relationship, a factor that was to lie behind the subsequent price jumps.

As inflation increased in the industrial countries at rates greater than were contemplated in the Teheran-Tripoli agreement and as the dollar was twice devalued, the OPEC countries complained that the escalation provisions of the 1971 agreements were inadequate. New agreements were signed at Geneva in January 1972 and June 1973, raising the posted prices by an aggregate of $0.35 a barrel. Together with the annual escalations previously agreed to at Teheran, these steps raised the posted price of August 1973 to $0.90 above the immediate post-Teheran level of $2.18 and the Saudi take to $1.81.

By September 1973, the OPEC countries no longer looked upon the Teheran agreement as satisfactory. OPEC was born of an effort on the part of the producing countries to prevent the posted price (the tax-reference price) from following the market price downward. The new organization succeeded in keeping the posted price at $1.80 during the market-price decline of the sixties. But in 1972, and more especially in 1973, the market price rose above the posted price, and the OPEC countries did not like to see the companies' profits getting the entire benefit of the price rise. Bargaining was accordingly renewed in October 1973, but negotiations were suspended at the outbreak of the 1973 Arab-Israeli war. A production cutback was instituted during the war by some of the Arab producers in order to enhance their embargo of the United States and the Netherlands. The cutback graphically demonstrated how high oil prices could climb when supply was restricted. In a series of unilateral moves, the Persian Gulf producers hiked the posted price from $3.01 on October 1, 1973, to $11.65 in January 1974. The Saudi take on light Arabian crude was accordingly raised from $3.05 to $7.00. Meanwhile, the Saudis had acquired 60 percent of Aramco, which meant that 60 percent of the oil output would be theirs to sell (almost exclusively to Aramco) at 93 percent of the posted price. The result was that the average revenue per barrel was raised—not to $7.00 but to $9.31. In January of 1975, Saudi Arabia made a pretense of reducing the price of oil by reducing the price to $11.25 while simultaneously altering other terms of the concession, so raising the average take to $10.12. This level of take was maintained through September 1975, when it was raised to $11.17 and the marker price to $11.51.

In the immediate aftermath of Teheran-Tripoli, the producing countries' attention focused on increasing their ownership and management of the oil companies' operations in their respective territories. In January 1974, Saudi Arabia acquired 60 percent of Aramco, and arrangements are now being made to acquire the balance. The role as offtakers of the international oil companies who are Aramco's stockholders will probably not be significantly altered, however. They will presumably still buy their oil production at cost, plus the government take. Most of the other OPEC countries have moved, or are planning to move, toward total ownership of the oil-producing operations in their territories.

THE FUTURE PROSPECTS OF THE CARTEL

Some analysts argue that for the oil companies to relinquish production activities in OPEC countries and become exclusively buyers of oil, continuing to play their present roles in transportation, refining, and marketing, would exert downward pressure on oil prices. As these large and competent buyers scour the world for the best bargains, each producing country would be encouraged to cut its price to make additional sales. As the prices fell, each producing country would be proportionately more anxious to maintain its sales volume, cutting its price even further to do so.

This argument delineates a possible scenario but not a necessary one. It is quite possible that the producing countries would continue to recognize their common interest in maintaining price and so resist the temptation to chisel. Much depends on the manner in which production comes to be allocated. Right now, the allocations effectively depend on how much the offtaking companies can sell. That does render plausible the idea that the companies are an important link in the chain of OPEC control of price. But, more fundamentally, how much the companies can sell at any time depends on the price at which they can afford to sell, which closely depends on the level of government take they have to pay. So right now, the allocation of sales is fundamentally governed by the pattern of takes.

In analyzing the future prospects of the cartel, it is helpful to distinguish between the pattern of takes and the level of take. The level may be identified with the Saudi take on marker crude (light Arabian 34°). The pattern may be specified by the set of differentials of the prices of all the other crudes from the marker crude. The future survival of the cartel depends upon whether a pattern of takes can be maintained that is consistent with a satisfactory pattern of outputs. In this connection, "satisfactory" applies primarily to Saudi Arabia. With a satisfactory pattern of differentials, there is a wide range of price *levels* that could be maintained in the future. If the level is to break, it will be because the pattern will have become unsatisfactory.

If Saudi Arabia were to be a residual supplier, with all the other OPEC countries producing all they wanted to, it is quite likely that Saudi Arabia would find the situation intolerable well before 1980. One analysis indicates that, if Saudi Arabia were to fulfill that role in 1980, its output would have to shrink to less than 2 m.b.d. as compared with a 1974 production of 8.5 m.b.d., a production capacity over 11 m.b.d., and a 1980 capacity projected before the 1973–74 price rise of 10 m.b.d. (see Table 4). Clearly, Saudi Arabia would find such an outcome unsatisfactory. But other outcomes are possible. In particular, if the "expansionist" OPEC nations in 1980 accept their 1973 output as their share and set their take accordingly, then Saudi Arabia's sales would also be just about the same in 1980 as in 1973.

From the OPEC viewpoint, the critical factor illustrated by Table 4 is, accordingly, how much the expansionist countries will increase their output. Of course, if the non-OPEC countries' demand grew by an additional 6 m.b.d. or so, there would be no serious problem. But OPEC can increase demand only by reducing the price of oil, thus giving the game away. But the rate of production increase by OPEC members *is* an

Table 4   Scenario for 1980, with Saudi Arabia as Residual Supplier

	1973 (M.B.D.)	1980 (M.B.D.)
World demand	57.5	77.3
Less non-OPEC production	26.6	46.5
USA (incl. natural gas liquids)	11.2	12.0
North Sea	0.0	6.4
Other non-OPEC	15.4	28.1
Equals demand for OPEC oil	30.9	30.8
Less Expansionist OPEC Production	12.4	17.9
Iran	5.9	8.0
Iraq	2.0	4.0
Nigeria	2.1	2.8
Algeria	1.1	1.1
Equals demand for conservative OPEC oil	18.5	12.9
Less non-Saudi conservative production	11.0	11.0
Venezuela	3.4	3.4
Kuwait	3.0	3.0
Libya	2.2	2.2
Abu Dhabi	1.3	1.3
Other Persian Gulf	1.2	1.2
Equals Saudi Arabia's residual share	7.6	1.9

Source: M. A. Adelman and Soren Friis, *Energy Policy*, December 1974, p. 288.

Assumptions:

1. World consumption rises at 7.5 percent per year. Elasticity of demand is −0.15, so that the 66 percent demand increase, 1973–80, is reduced to 34 percent in response to the price increase to $9 (in 1974 dollars).
2. Other non-OPEC production has a supply elasticity of 0.35.
3. Expansionist OPEC countries' production is going to rise as shown in table.
4. Non-Saudi conservative OPEC countries' 1980 production will equal 1973.
5. Saudi Arabia will get what is left. (The purpose of the exercise is to see what would be left given the above assumptions.)

OPEC AFFAIR. Their critical problem then is to see that the output of these countries grows by an amount no more than the conservative countries are prepared to let their outputs shrink.

There are many possibilities for this projection to go wrong, of course. The OECD, for example, estimates that at a $10.50 price (in 1974 dollars)[13] OPEC sales in 1980 would run at 25.7 m.b.d., thus throwing the burden of another 5 m.b.d. cutback onto OPEC. If that additional cut were to be concentrated on the Persian Gulf countries and Libya, it would reduce their 1980 output to 4.4 m.b.d., or less than one-third of their 1973 output of 15.2 m.b.d. That would quite clearly be unacceptable to them.

The Federal Energy Administration (FEA) estimates that at $9.92 (in 1974 dollars) per barrel, OPEC sales in 1985 would be 28.3 m.b.d.[14] But whatever the demand turns out to be, survival of the cartel depends upon the allocation of output remaining satisfactory to the participants. The experience of 1974 and the first three quarters of 1975 suggests that there can be huge asymmetries of cutbacks without straining the cartel's cohesion (see Table 1). Ample evidence exists that most of the producing countries are prepared to cut back output approximately in proportion to their production capacity. The principal exception is Iraq, which for almost its entire history has had its progress as an oil producer blunted for one political reason or another. But even Iraq has reduced its 1980 production target from 6 m.b.d. to 4 m.b.d. As more OPEC members increase import expenditures up to or above the flow of receipts from oil, their readiness to accept cutbacks may be weakened. Their concern for maintaining the price may be strengthened, it is true, but that is a general concern that cannot be expected to stand up against the particular concern of not taking a loss in volume. On the outcome of this tension, the life of the cartel will depend.

Sometime in the 1980s, it may be hoped, production outside OPEC may grow so large as to require such extensive cutbacks by the OPEC countries that some of them may start either to cheat or reduce their takes openly, or both. When this tendency begins, it can be expected to snowball. The compulsion to maintain revenues will probably lead to competitive adjustments of take differentials that will cumulate on each other to accelerate the price decline. The price can be kept high only by universal restraint of output. Once the price starts declining, two sorts of effort can be expected. One involves joint moves to reconstitute the cartel. The other involves separate action on the part of each to increase his sales at the expense of others. If, as is likely, there is a lot of oil in the world to be discovered and produced while the price of $11.51 a barrel (1975 dollars) lasts, the tendency to cut price will ultimately prevail. But that will not be soon. For the next five years at least, only an internal change of heart, hardly to be expected, could break the OPEC price. This judgment sets aside the possibility, always with us, that another round of the Arab-Israeli war may break out, with incalculable ultimate effects on the world oil supply in general and on the oil price in particular. In the event of another war, the Arab oil producers would doubtless be under very strong pressure to institute another embargo and cutback. The market price of oil certainly could be expected to rise, but from that point on it is difficult, if not impossible, to predict what would happen. One observation is worth making, however: if the oil supply is manipulated for political and military purposes, the future of OPEC may be subject to violent change.

Short of such a development, the life of OPEC and its high price level will depend primarily on the level of demand for OPEC oil at that high price relative to the revenue requirements of the producers. The most significant factor affecting the demand for OPEC oil is the production of oil in the non-OPEC countries. This in turn hinges upon the rate of exploration and development in these countries; in particular, the 82.5 percent increase in the oil production of other non-United States countries in Table 3 is a highly speculative estimate. These countries possess a very large proportion of the world's sedimentary deposits, but those deposits have not, on the whole, been thoroughly explored. In any case, a major contribution from them beyond the estimate in Table 3 could be expected only after 1980. Conservation on the part of the importing countries as well as the development of alternative fuels will also play a role but are likely to remain of secondary importance.

CUTBACKS IN DEMAND

The conservation targets, sought by the United States in its dealings with other consuming countries, were set at 2 m.b.d. by the end of 1975 and 4 m.b.d. by the end of 1977, with the United States accounting for half of each target and the 17 other members of the International Energy Agency (IEA) for the other half.[15] There is considerable doubt that these goals will be achieved. There will, of course, be "movement along the demand curve," that is, a lower level of consumption because of the high prices. But it is doubtful that there will be a significant additional "shift of the demand curve," so that by national policy a substantially smaller amount of oil will be consumed in each country than would have been consumed at the new high price under the old demand conditions.

Whether conservation cutbacks in consumption on the scale of 2 to 4 m.b.d. can be achieved or not, there is some question of how great an influence the cutbacks can exert on price. Skeptics in and out of Congress asked, when the cutbacks were proposed, if OPEC's then current overcapacity of 8 m.b.d. was not causing OPEC to crumble, then why should an overcapacity of 10 m.b.d. or even 12 m.b.d.? The recession demonstrated the validity of the argument, for the higher levels of overcapacity were reached without any noticeable rents in OPEC's fabric. As general economic recovery proceeds, world oil consumption will share in that recovery, and the higher consumption and the termination of destocking will reduce the overcapacity. Any further progress in cutting demand is likely, in the near future, to be less effective in raising surplus capacity than was the recession. Unless greatly augmented by increased non-OPEC production of oil and oil substitutes, consumption cutbacks alone are unlikely to affect OPEC's price critically.

Professor Adelman maintains that if the oil-importing countries cut back consumption, they are issuing an invitation to the producing countries to raise their price.[16] This paradoxical argument hinges on the assumption that the cartel is maximizing its revenues at a relatively elastic point on its demand curve and that a consumption cutback would curtail the less urgent uses, thus propelling the market to an inelastic point on the demand curve.

An OPEC price rise would be profitable if the cutback involved a quantitative restriction, such as an import quota. If, for example, a quota was gradually imposed by the United States, ultimately cutting its imports by 2 m.b.d., and a proportionately equal cutback was adopted by the other OECD countries, the importing countries' domestic price of crude could be expected to rise by at least $4 a barrel. If, after the consumer-country cutbacks, OPEC raised its price by $4 to $15.51 at the Gulf, its members would lose no sales dollars but rather would eliminate the otherwise unsatisfied demand created by the OECD quotas at the old price. By its price rise, OPEC would achieve, through the effect on demand, the same import cutbacks assumed to be effected by the quotas; therefore, the quotas no longer would be effective constraints. If the OECD countries, recognizing this possibility, achieved a cutback by imposing a $4 import tariff, preempting the proceeds of the price rise, Professor Adelman argues that OPEC would still gain by increasing prices even further, since the demand would be more inelastic after the cutbacks. That is more questionable.

The argument presumes that OPEC is a monopoly maximizing its profits at the $11.51 price. I do not believe this to be the case. High as the OPEC price is, a solidly entrenched monopolist could profitably charge a higher price in the short run. But in OPEC's case, the monopoly may not be quite so perfect, and the output cutbacks required at the higher prices might put an undue strain on the cohesion of the cartel.

In the long run, the higher prices would stimulate non-OPEC output of oil and oil substitutes. But that day is still in the very distant future, and many obstacles must be overcome by the non-OPEC countries before their domestic supplies of oil and oil substitutes compete with OPEC's price-setting powers.

How much an import cutback would harm the OECD countries is a major question. Suppose, for example, that a $4 tariff results in a cutback of 4 m.b.d. for all OECD members. If OPEC lowered its netback price to $7.50 to prevent its excess capacity from rising, the move by OECD would be a great success. Specifically, OECD members would enjoy a net gain of about $44 billion a year. If, however, OPEC stood fast with a $11.50 price, the OECD countries would suffer substantially. Each of the 4 m.b.d. sacrificed would be worth from $11.51 to $15.51 to the would-be purchasers, since they represent the barrels not purchased once the price to the buyer rises from the first to the second level. The average value per barrel foregone is $2 multiplied by 4 m.b.d. and 365 days in a year. The tariff would thus cost $2.9 billion.

On the other hand, FEA estimates, though it is extremely difficult to accept, that the cost of an oil cutback to the United States would amount to $33 billion a year per m.b.d. sacrificed after the first[17]—the first daily million are deemed costless because of immediately available economies. On this basis, an equal cutback by other OECD members would presumably cost about the same.

A NEGOTIATED PRICE CUT?

The FEA estimates of cutback costs are quite possibly excessive. But the hope that a $4 OECD tariff on crude would drive OPEC's price down in the medium run also seems overly optimistic.

On the other hand, if at a meeting between producers and consumers it was agreed to rescind the consumption cutback, and in particular the $4 tariff, in exchange, say, for a $2.50 reduction in the price of oil to $9 at the Gulf, there might be a net gain all around, except from the viewpoint that it is meritorious to consume less oil for its own sake. This outcome would yield about the same return to OPEC as would the preceding variant that ended up with a cut of 4 m.b.d. at an $11.51 price plus a $4 tariff. If, for example, OPEC could sell 32 m.b.d. at $9 and no tariff, it would net slightly more than from selling 24 m.b.d. at $11.51 with a $4 tariff. And overcapacity would be reduced by 8 m.b.d. at the lower price.

It is not likely that OPEC would be willing to cut its price so substantially in exchange for a reduction of a tariff that would otherwise be imposed. But there is no question of the validity of the general argument underlying the preceding example, although the numbers used are only guesses. For any OECD tariff of X there is clearly a price reduction of Y,

such that OPEC will earn as much at a price of $11.51 minus
Y and no tariff as at a price of $11.51 with the tariff.

If OPEC was led to believe that unless it cut its price by Y,
the OECD would impose a tariff of X, then OPEC might
negotiate a price reduction of Y in return for a promise of no
tariff increase. But the higher X is set, the more damage will
be self-inflicted on the OECD economies by the oil-
consumption cutback if OPEC stands firm at $11.51. By re-
fusing to tender any substantial price cut, OPEC may shake
OECD from its strategy of trading a tariff reduction for a
price cut. The argument that OPEC would be willing to trade
a price cut for the removal of a tariff on imports also assumes
that the "user cost" of the extra 8 m.b.d. of output is out-
weighed by considerations of surplus capacity. That user cost
is the sacrifice of the opportunity to sell that oil some forty
years from now at a real price of about $15 per barrel dis-
counted to the present at, say, 4 percent.

If OPEC members accept a lower price, it will be because
they fear that the cartel would be weakened otherwise and the
price would sink even lower. Under such circumstances a
negotiated agreement should not be attractive to the importing countries.

It is frequently said that the importing countries can gain
"assured supplies" from a commodity agreement. The phrase
seems to be an empty one, however. Either normal commercial channels are open and anyone who can match the bids of
other customers can buy oil, or there is an embargo with
cutbacks and all customers are forced to pay more. An embargo without cutbacks is an empty gesture. The embargoed
countries will get their oil from the producers not party to the
embargo or from the customers of those who are. While some
producers may grant some importers special assurances of
supplies in an emergency, these assurances are likely to be of
doubtful value when an emergency arises. The total thrust of
the International Energy Agency is to unify the principal
importing countries against a selective embargo. The prece-
dent of the 1973 embargo cutback is not encouraging. Article
34 of the Treaty of Rome, which was supposed to have uni-

fied the European Economic Community (EEC) countries' access to raw materials, was not honored in that case.

INCREASED PRODUCTION OF OIL AND SUBSTITUTES

The prospects are not good that a voluntary cutback in the consuming countries will break the OPEC price. Nor are they any better that the increased production of oil substitutes will, in the next 5 or 10 years, achieve that end. For 1980, the commitments would have to have been made already—and they have not been. Even for 1985, syncrude or nuclear power will be forthcoming only from plants that will have been started by 1976.

Over a longer horizon, cost will limit the increased production and use of oil substitutes. The use of coal as underboiler fuel is limited by the expense of conversion of the burner; its use as the basis of synthetic oil is limited by the expense of synthesis, principally the cost of the capital required. Even over the long run, with coal replacing oil and gas in underboiler use, an OPEC output of 28 to 30 m.b.d. will be required. It appears that the United States equivalent of the $11.51 Gulf price of oil (about $12.50 at 1975 prices) may be inadequate to cover the costs of most syncrudes—oil derived from coal or shale. But the same $11.51 price, plus transport advantages outside the Middle East, should eventually cover the costs of finding, developing, and producing crude oil throughout the world. If there were holes in the ground throughout the world in the same density relative to sedimentary deposits as there are in the United States, the potential supply of crude oil would be huge enough to drive the price down to the competitive level.

But that is not how things stand now. Drilling activity in the rest of the world is far less intense than in the United States (see Table 5). This reflects the great disparity in oil-productive capacity attained per foot drilled outside the United States as compared to our current capacity. The OPEC countries sustain an output three and one-half times as large as that of the United States and an incomparably larger

Table 5   Drilling Activity, Production and Reserves in OPEC and Non-OPEC Countries

	(1) FOOTAGE DRILLED, 1973 TOTAL (MILLIONS OF FEET)	*(2)* WILDCAT FOOTAGE, 1973 (MILLIONS OF FEET)	*(3)* PRODUCTION, 1974 (M.B.D.)	*(4)* RESERVES JAN 1, 1974
Non-OPEC Countries	172.6	60.2	14.5*	127.6
United States	138.9	44.8	8.8*	35.3
Canada	16.8	8.4	1.7	9.4
Latin America	10.0	2.9	1.8	11.9
Europe	3.6	2.1	0.4	16.0
Middle East	0.7	0.1	0.5	37.2
Asia Pacific	1.6	1.4	0.9	5.1
Africa	1.0	0.6	0.4	12.7
OPEC Countries	14.5	3.0	31.1	396.3
Middle East	4.4	0.4	21.7	311.5
Africa	4.0	1.0	4.9	54.6
Asia Pacific	2.3	1.0	1.4	10.5
Latin America	3.8	0.6	3.1	19.7
World (Non-Communist)	187.1	63.2	45.6*	524.0

*Excluding natural-gas liquids

Source: *International Petroleum Encyclopedia,* 1973, 1974, 1975

growth of reserves on a level of drilling activity only one-tenth as much as that of the United States. Indeed, the entire non-Communist world, outside the United States and Canada and the OPEC countries, accounts for only one-eighth as much footage drilled as does the United States. Oil found and developed will, in any area, be roughly proportional to footage drilled. The factor of proportionality varies greatly from area to area, however. An onshore well at an average location outside the United States (even excluding the Middle East) will produce immensely more per day than a United States well. The most immediately practical way of increasing the world's oil-productive capacity would be to shift drilling resources from the United States to other countries or at least to augment exploration and development activities outside the United States. This consideration applies most particularly to onshore wells. Offshore prospects in waters adjoining the

United States are not so much different from those in other localities.

A powerful brake is currently operating on the rate of exploration and development. The great price rise of oil is so recent and has taken such a form that its transmission into an incentive to explore for oil and to develop what is found is imperfect and ineffective. The high prices go almost entirely into the government take. That leaves, as an incentive to the private company that must take the risk, only a very small part of the price. Ultimately, the non-OPEC governments will come to recognize the situation and take the appropriate action. But they have not done so as yet.

As the principal beneficiaries of any oil or gas discovery, the sovereign governments will find it to their advantage to stimulate exploration and development. This they can do by taking a major share of the risk, using any one of a number of devices—tax allowances, subsidies, joint ventures, etc. Even prior to the overt taking of such risks would be a reconsideration of the current legal provisions with the view to eliminating the perverse incentives. Oil exploration is reported not to be rig-bound any longer, as it should certainly be expected to be in view of the jump in price and the long lead time required for delivery of a new rig. The only explanation would seem to be that exploration and development are being inhibited by institutional conditions—principally tax structures outside the United States and leasing terms both in the United States and abroad.

A COMPETITIVE BENCHMARK FOR THE PRICE OF OIL

If OPEC collapsed because some of its members insisted on expanding production and because the price of oil became competitive, it would be between $1 and $4 a barrel, depending on the rate of discount. The competitive equilibrium price of a barrel of oil must cover not only the cost of production but an allowance—called "user cost"—for the loss of one barrel of oil from the reservoir. The user cost of a barrel of oil at a particular time can be approximately estimated as the future price to be expected when the last barrel is exhausted, discounted back to the time under consideration.[18] That ultimate future price can be expected to be equal to the "backup price," that is, the cost of the closest substitute that will take over the market at that time, with allowance for any quality differential. Suppose, for example, oil were competitively produced and is confidently expected to run out in 40 years, at which time the closest substitute would be shale oil or a coal-based synthetic produced at a cost of $15 a barrel in 1975 dollars. And suppose further that the appropriate real rate of interest (the nominal rate minus the rate of change in the price level) at which to discount the future price is 4 percent. Then the user cost of a barrel of oil would be $\frac{\$15}{(1.04)}$ 40, or $3.13. Under these assumptions, the competitive price for Middle Eastern oil, with a production cost of $0.17 a barrel would be about $3.30 in 1975 dollars.

If the appropriate discount rate is 2 percent, the user cost, still assuming the same reservoir life and backup price, would be $6.79, the corresponding competitive price about $7. On the other hand, if the appropriate discount rate is 8 percent, the user cost, based on the above assumptions, would be $0.69 and the competitive price less than $1.

Under present conditions, with the Persian Gulf price at $11.51, it may seem unrealistic to talk seriously of a price under $1. Yet in the 1950s and 1960s, the price trended in that direction. User cost did not seem to enter into the decision-makers' calculations.

What discount rate is in fact appropriate? That rate, d, which would balance the marginal producer's decision between producing and not producing another barrel if next year's price, P_{t+1}, is expected to equal $P_t(1 + d)$. Were OPEC to crumble, the effective discount rate might rise very high and the user cost sink close to zero. For if the price of oil were to break sharply and deeply, most producers would want to produce at full capacity in order to meet foreign expenditure commitments, even if next year's price were expected to exceed this year's by a large margin. Under such circumstances, the price of oil would be driven down below the direct cost of production of the higher-cost producers until enough production capacity was knocked out in order for a new equilibrium to be attained.

Spokesmen for the oil producers, the Shah of Iran in particular, frequently justify the price set by OPEC in terms of the cost of the closest substitutes. That is indeed an appropriate justification for a monopoly, which can maximize its profits by raising the price to the point where demand ceases to be inelastic. The initial point for crude oil at which the price becomes elastic in the long run is the backup price, the cost of the cheapest equivalent substitute in elastic supply. This may be placed in the neighborhood of $12 to $15 at 1975 prices in the United States for shale oil from high-grade deposits (25 gallons to the ton or more). With allowances for transportation, the current Persian Gulf price, at $11.51, comes close.

Under competitive conditions, the price of oil can be expected eventually to approach the cost of the closest equivalent substitute, but not now. To satisfy the competitive price standard, the current price of oil should equal not the current *cost* but the *discounted* cost of the oil substitute or replacement. Such a discounted cost will be highly sensitive to the length of the period before the oil is exhausted, as well as to the discount rate. The time frame to be applied is not the current proved reserves divided by current output but the time it will take expected future sales at the competitive price to exhaust current reserves plus expected future discoveries.

UNITED STATES PRICE POLICY AND PRICE PROSPECTS

In the first days of the major upswing of prices immediately after October 1973, the problem of the embargo-cutback loomed so large that it completely overshadowed the price rise. The price rise tended to be regarded simply as the consequence of the cutback, presumably to be reversed when the embargo-cutback was terminated. The main thrust of United States foreign oil policy, therefore, from October 1973 to March 1974, was to get the embargo lifted and, more importantly, to have the cutbacks rescinded.

When the embargo-cutback was finally terminated in March 1974, the problem of the immediate future was no longer the supply of oil—there was plenty forthcoming—but the price of oil. Some of the importers, particularly Western Europe and Japan, were, however, still preoccupied with supply.

Early in 1973, there were suggestions in the U.S. State Department for an international organization of consuming nations, primarily for "international allocation of the available oil if chronic shortages occur."[19] The embargo-cutback gave impetus to these plans. A few weeks after President Nixon had proclaimed "Project Independence" in Washington in November 1973, Secretary Kissinger delivered a speech to the Pilgrim Society in London calling for the Atlantic allies to set up an "energy action group" to develop, within three months, an action program for "collaboration in all areas of the energy problem."[20] The leading countries of Western Europe, and Japan as well, received the suggestion coolly—they wanted nothing that would suggest confrontation with the oil suppliers. They did, however, attend a 13 nation energy conference in Washington in February 1974.

Out of this meeting there developed in November 1974, after many vicissitudes, the International Energy Agency, under the auspices of the OECD, and the abortive Paris conference of oil producers and consumers of April 1975. The IEA has 18 member countries, including the United States, Japan, and the nations of Western Europe, with the exception of France, which participates as an observer. The underlying philosophy of the United States in sponsoring the IEA is that only through "a common and comprehensive approach by consumer countries to energy problems . . . can we hope to solve the world energy crisis."[21] The IEA adopted the International Energy Program (IEP), under which the member countries undertake three commitments subject to government ratification:

> 1.   To build common levels of energy reserves (tentatively set at a 90-day supply but not yet generally implemented)

2. To develop demand-restraint programs that would enable them to cut oil consumption by a common rate in an emergency

3. To allocate available oil in an emergency

Critics raise two important questions about the effectiveness of their program. First, under a renewed embargo-cutback the same "devil-take-the-hindmost" spirit would probably develop, as in 1973–74, despite the agreement. Second, even if the agreement is honored, it puts the potential embargoer in the same position as in 1973. The 1967 experience taught that an embargo without a cutback is ineffective. The 1973 experience taught that the effectiveness of an embargo-cutback depends primarily on the extent of the cutback. Cutting off oil from certain consuming countries merely opens the way for other countries to supply the cutoff countries if world output is not cut back.

The IEP aims have no bearing on the price of oil, other than forging greater cohesiveness among the consuming countries. Such consumer cohesiveness as has been achieved seems to have limited leverage on price determination. The oft-proposed substantive meeting of the consumer countries with the producer countries, now scheduled for December 1975, may have a bearing on price. Early in 1974, Saudi Arabia and Iran declared their readiness to work cooperatively with the consuming countries on the price of oil. A second compromise between the French desire for a producer-consumer conference and the American desire for a united front of consumer countries was announced at the Martinique meeting of Presidents Ford and Giscard-D'Estaing in December 1974, substantially echoing the Washington Conference of the preceding February.[22] A preparatory conference at Paris in April 1975 was dominated by the demands of the Third World, voiced primarily by Algeria, for a "new economic order," which would rearrange the basis of payments for raw materials by developed countries. With the United States insisting that it was to be an energy conference and with the OPEC countries supporting Algeria, the conference broke up without any agreement.

The United States gradually modified its position until a new stand was announced in a speech delivered, on behalf of Secretary Kissinger, by Ambassador Moynihan at the UN. It was widely reported to have involved, in its gestation, strong conflicts within the administration, especially between the Secretaries of State and Treasury. Both men stressed that the message was "developed jointly."[23] Yet it is hard to believe that Treasury's espousal of the operation of free markets (reaffirmed in the speech) could be consistent with the likely outcomes of "consumer-producer forums for every commodity."

The United States put itself on the record as advocating a number of concrete measures for helping the developing countries. In the special UN session of September 1975, devoted to economic development and international cooperation, experienced observers claimed to note "a new air of conciliation" and a "retreat from earlier third-world rhetoric."[24] Thus, the participants are looking forward to the conference "on energy and other problems of the rich and poor nations," scheduled for December 15, 1975. The eventual substantive conference, it is planned, will be divided into four separate commissions: energy, other commodities, development problems, and financial questions, respectively. These are to proceed in parallel—an expression whose ambiguity remains unresolved—for the United States position is that the four topics are separate, while the Third World spokesmen, probably including the OPEC countries, regard them as linked. Clearly, negotiations on the price of oil are likely to proceed in a framework of negotiations on many other matters.

The moderate price use adopted at the September 1975 OPEC meeting has been attributed to the progress made on producer-consumer negotiations, although economic reasons probably were also important. But in the intermediate and long run, it will be the relationship of the supplies of oil to the market that will govern the price of oil, and here the lack of progress has been discouraging. The North Sea development appears to be slowing down, relative to earlier expectations, largely because of the projected United Kingdom take. Alaskan development may have been slowed down by the delay in deliveries due to the prospect of the failure of a "window" to open up on time in the ice jam in the summer of 1975. Nuclear projects are being delayed or canceled. Syncrude plans are simply not being developed. Exploratory rigs are not being fully utilized, nor is the rig-building capacity. Leasing is still lagging.

Price reduction through producer-consumer conferences is still possible in theory, but it is hard to see how it can come about. Sheik Yamani, said to be the principal architect of Saudi oil policy, has declared: "Lower the prices of the goods you sell us, which you can since you have a monopoly in the field, and we'll behave accordingly."[25] Aside from the confusion between a monopolistic seller and a group of competing sellers who jointly exhaust the market, the general attitude foretells a negotiating stance aimed at indexation. The OPEC countries will say, as Yamani did elsewhere, you keep your product prices steady and we'll keep the oil price steady. The importing countries would appear to have the choice of stabilizing their own prices or of seeing the nominal price of oil rise with their own price levels. In either case, the outlook is for a real price of oil close to its present level as long as OPEC remains cohesive.

III/Consequences of the Price Rise

By 1976, world oil imports from OPEC can be expected to run at 30 m.b.d., as they did in 1973 and 1974. During the recession of 1974–75, world imports from OPEC fell as low as 25 m.b.d. They may average 27 m.b.d. for all of 1975. Even during the recession, however, stocks were being depleted, so that world consumption of OPEC oil was not much below the 30-m.b.d. figure.[1] When imports are running at 30 m.b.d., each dollar rise in the price of oil means a $30 million increase in the daily cost of oil to the importing countries, or about $11 billion a year. When imports are running at 27 m.b.d., each dollar in the price means a $10 billion difference in total annual cost.

The Saudi government take rose from $0.89 a barrel in 1970 to $10.17 in October 1975 (see Table 3). On the assumption that other countries' takes rose by an equal amount, the increase in total government takes, at 30 m.b.d., amounts to $113 billion in 1975 dollars per year. In 1975 dollars, the 1970 take was $1.25 per barrel, and the October 1975 take is figured at $10.96 (see Table 3). The increase in take from 1970 to October 1975 was, accordingly, $9.71 in 1975 dollars. The corresponding increase of the total annual rate of take on 30 m.b.d. has been $106 billion a year. At 27 m.b.d., the increase has been approximately $96 billion. The increase in aggregate take constituted most of the increase in costs to the consuming countries. The companies' margins do not seem to have changed by more than an insignificant amount as compared to the change in the rate of government take.

The impact of the price rise on developed and on less-developed countries is illustrated in Table 6. The OECD countries, accounting for over 85 percent of the world oil imports, have a corresponding share of the cost increase: $92 billion a year. The Fourth World's yearly share is much smaller, about $10 billion. The increased cost for them came to 1.5 percent of their 1973 GDP, and for the OECD countries 2.4 percent. In terms of GDP, the impact was hardest on Japan; relative to growth of income, on Western Europe.

Economists, analyzing the consequences of the oil price hike, employ measures like those in column 5 of Table 6, showing the cost increase as a percentage of GDP, or those of column 6, the increased cost as a ratio of the annual growth rate. Thus, increased costs were less than 2.5 percent of the GDPs of the developed countries and not more than 1.5 percent of the GDPs of the less-developed countries. Only six

Table 6   Impact of the Oil Price Rise

	(1) NET OIL IMPORTS, 1974 (MMB/D)	(2) IMPACT OF PRICE RISE ON IMPORTS (COL. 1 × $9.71)	(3) GROSS DOMESTIC PRODUCT (GDP) (1973 AT 1975 PRICES IN BILLIONS OF DOLLARS)	(4) GROWTH RATE (1960–72 AVG.)	(5) IMPORT COST RISE AS PERCENT OF GDP (COL. 2 ÷ COL. 3)	(6) IMPACT OF PRICE RISE IN MONTHS OF GROWTH
United States	5.9	20.9	1,590	4.1	1.3	3.8
Western Europe	14.4	51.0	1,591	5.1	3.2	7.5
Japan	5.4	19.1	491	11.0	3.9	4.2
Australasia	0.3	1.1	93	4.6	1.2	3.1
OECD*	26.0	92.1	3,765			
Non-OPEC LDC	2.8	9.9	661	6.1	1.5	2.9

*Excluding Canada, an oil exporter by a small margin (see Table 2).

Sources: Column (1); see Table 2

Column (2): column (1) × $9.71, the difference between the Oct. 1975 take ($10.96) and the 1970 take ($1.25), both in 1975 dollars (see Table 3).

Column (3): for OECD countries, *OECD Statistical Yearbook*, 1973 dollar figures converted to 1975 dollar figures by use of United States GNP implicit deflator.

Column (4): OECD figures from OECD; LDC figures from World Bank (LDC figures are for 1960–70).

months' growth of output, at the rates of the 1960s, would be required for the developed countries to absorb the extra costs of oil imports; and less than three months for the less-developed countries.

The use of such measures is frequently attacked by non-economists, or by anti-economists, on the grounds that these measures understate the full consequences of the cost increase. In one sense, the figures overstate it. The OPEC countries are not spending all of their foreign revenues on imports—the non-OPEC countries are therefore not being drained of their resources to the full extent indicated in Table 6. They pay for the balance of the increase on credit. For the non-OPEC countries as a group, the balance will equal the ratio of the OPEC surplus on current account to the total of OPEC exports—about 60 percent in 1974 and 40 percent in 1975.

THE ACCUMULATED SURPLUSES

The problems that have attracted a major share of attention are not those of the increased real transfers of goods and services extorted from the importing countries by the oil exporters. More attention has been paid to the financial aspects, particularly to the prospect of a huge "mountain of claims" to be accumulated by the OPEC countries as the cumulative counterpart of their surpluses on current balance.

A World Bank study projected accumulated foreign (OPEC) assets at $653 billion by 1980 and $1,206 billion by 1985 (see Table 7). (For details of two of the studies, see Tables 8 and 9.) These projections differed with respect to three principal factors:

1. The future volume of OPEC exports
2. The future price of oil
3. The future levels of OPEC imports

The projections generally are the same with regard to the level of OPEC exports as trending, with some fluctuations, from the depressed levels of 1975, toward 30 m.b.d. by 1980. Prices, the second factor, are subject to wider differences. For

Table 7 Various Projections of OPEC Surplus Accumulation

PROJECTOR	DATE PUBLISHED	CURRENT DOLLARS (BILLIONS) 1980	1985	CONSTANT 1974 DOLLARS (BILLIONS) 1980	1985
World Bank	July 1974	653	1,206	411	519
OECD	July 1974	—	—	250–300	—
OECD	Oct. 1974	"up to 500"	—	250–325	—
World Bank (Chenery)	Jan. 1975	—	—	"up to 300"	—
Morgan Guaranty	Jan. 1975	179	—	—	—
U.S. Treasury (Willet)	Jan. 1975	—	—	{ 200–300 / 200–250	—
Irving Trust	Mar. 1975	22–248	—	—	—
World Bank	April 1975	460	—	248	—
"Middle East Producers"	April 1975	—	—	100–125	—
U.S. Treasury (Parsky)	April 1975	—	—	250	—
U.S. Treasury (Parsky)	May 1975	—	—	200–250	—
Citibank	June 1975	189	30	—	—
W. J. Levy	June 1975	449	—	286	—
OECD	July 1975	—	—	215	200

Sources:

World Bank, July 1974: *Staff Report 477*, July 8, 1974.

OECD, July 1974: OECD, *Economic Outlook*, no. 15, July 1974, p. 95.

OECD, October 1974: OECD, *Energy Prospects to 1985*, Paris, 1974, v. 1, p. 70.

World Bank (Chenery), January 1975: Hollis Chenery, *Foreign Affairs*, January 1975, p. 254.

Morgan Guaranty, January 1975: *World Financial Markets*, January 21, 1975, p. 8.

U.S. Treasury (Willett), January 1975: "The Oil Transfer Problem," basis of a speech by Thomas D. Willett, Deputy Assistant Secretary of the Treasury for Research, *Dept. of the Treasury News* (press release), January 30, 1975.

Irving Trust, March 1975: *The Economic View from One Wall Street*, March 20, 1975.

World Bank, April 1975: Staff study quoted in *Middle East Economic Survey*, April 25, 1975.

"Middle East Producers," April 1975: Quoted by Assistant Secretary of the Treasury Parsky, *Middle East Economic Survey*, April 25, 1975.

U.S. Treasury (Parsky), April 1975 (same as previous source).

————, May 1975: *The Wall Street Journal*, May 23, 1975.

Citibank, June 1975: First National City Bank, *Monthly Economic Letter*, June 1975.

W. J. Levy, June 1975: W. J. Levy Consultants Corp., "Future OPEC Accumulation of Oil Money: A New Look at a Critical Problem," report of June 1975.

OECD, July 1975: OECD, *Economic Outlook*, no. 17, July 1975, p. 79.

Table 8   The Morgan Guaranty Scenario

Assumptions:

1.   Little change in oil exported.
2.   A 5 percent annual increase in OPEC government take per barrel. (This leaves unexplained the smaller increase in revenues in 1978.)
3.   An average 15 percent annual increase in OPEC revenues from non-oil exports.
4.   A 20 percent increase in the *volume* of OPEC imports.
5.   An average 7 percent annual increase in the *price* of OPEC imports, except 12 percent in 1975.
6.   An 8 percent annual return on OPEC external investments.

	1974	1975	1976	1977	1978	1979	1980
			(BILLIONS OF CURRENT DOLLARS)				
OPEC exports of goods and services	112	117	127	135	139	148	158
Oil revenues	105*	110	119	125	128	135	143
Non-oil exports	7	7	8	10	11	13	15
OPEC imports of goods and services	50	65	83	108	138	177	227
Trade balance	62	52	44	27	1	−29	−69
Investment income	3	8	13	16	19	19	16
Current account	65	60	57	43	20	−10	−53
Grant aid	2	3	3	3	3	3	3
Surplus to be invested	63	57	54	40	17	−13	−56
External financial assets†	80	137	191	231	248	235	179

*Value of oil exported. Exceeded actual revenue receipts by an estimated $10 billion.
†Cumulative amount outstanding at year-end.

Source: Morgan Guaranty Trust Company, *World Financial Markets*, January 1, 1975, p.8.

example, the projection published by the Morgan Guaranty Trust Company assumed a 5 percent annual rise in oil prices and overall inflation of 7 percent, resulting in a 1980 price of $13.25 per barrel. Walter J. Levy assumed a price rise of 12 percent in 1976 and of 7 percent thereafter, leading to a $14.65 price in 1980. First National City Bank of New York projects a low level of accumulation, based principally on the assumption of a price decline to $9.10 by 1980. The Morgan and Levy projections differ most widely in their estimate of OPEC imports of goods and services—the 1980 level is pegged at $227 billion and $164 billion, respectively (4.5 and 3.6 times 1974 imports).

The multiplicity of projections vividly illustrates the uncertainties of the situation—particularly with respect to the price of oil and the rate of increase of imports of the OPEC countries. At the moment, it appears that the future price of oil will do no more than keep up with the rate of inflation, while the increase in imports by the OPEC countries has surprised many, if not all, observers (see Tables 10, 11, and 12). The recent past would seem to lend credence to the Morgan projection. Walter J. Levy argues that the oil price will rise further when the importing countries' economies recover and the OPEC imports encounter bottlenecks.

Table 9   The Walter J. Levy Scenario

Assumptions:

1. Volume of OPEC oil exports to recover to 30 m.b.d. in 1976 and then rise and fall back.
2. A 7 percent annual increase in OPEC government take per barrel, except for a 12 percent increase in 1976.
3. OPEC non-oil export revenues to increase from $7 billion in 1974 to $19 billion in 1980, "in line with Morgan scenario." (Actually comes to $15 billion more, taken cumulatively.)
4. A 15 percent annual increase in the *volume* of OPEC imports.
5. Same assumption of *price* increase of OPEC imports as in Morgan scenario.

6. A 7 percent annual return in OPEC external investments.

	1974	1975	1976	1977	1978	1979	1980
	(BILLIONS OF CURRENT DOLLARS)						
OPEC oil exports (m.b.d.)	29.6	26.5	30.0	31.5	32.5	32.5	31.5
(bil. bbls. per year)	10.80	9.67	10.95	11.50	11.86	11.86	11.50
Per-barrel revenues ($/bbl.)	9.72	10.00	11.20	12.00	12.80	13.70	14.65
OPEC total oil revenues	95*	93*	123	138	152	162	168
Non-oil exports	7	8	10	12	14	16	19
Total revenues	102	101	133	150	166	178	187
Imports of Goods and Services	45	58	71	88	108	133	164
Balance of trade	57	43	62	62	58	45	23
Investment Income	3	7	11	16	21	26	30
Grant aid (−)	2	3	4	5	6	6	6
OPEC current surplus	58	47	69	73	73	65	47
OPEC cumulative surplus (year-end)	75	122	191	264	337	402	449
Adjustment to 1974 Dollars (assumed price index of OPEC imports, 1974 = 100)	100	112	120	128	137	147	157
OPEC current surplus deflated by price index (in 1974 dollars)	58	42	58	57	53	44	30
OPEC cumulative surplus deflated by price index (in 1974 dollars)	75	109	159	206	246	273	286

*Actual cash receipts. Assumes a payments lag of $10 billion in 1974 and a further $4 billion in 1975, with a payments lag of $14 billion per year remaining constant thereafter.

Source: W. J. Levy Consultants Corp., *Future Accumulation of Oil Money: A New Look at a Critical Problem*, New York, June 1975.

Table 10   OPEC Imports: Past and Projected

Merchandise Imports	1973	1974 (Estimated)	1975 (Projected)	1980	1985
		(BILLIONS OF 1974 DOLLARS)			
Algeria	2.1	3.7	5.7	6.5	10.0
Ecuador	0.5	0.8	0.9	1.5	2.2
Indonesia	2.4	3.9	4.7	9.4	12.3
Iran	3.6	8.0	10.6	24.4	32.0
Iraq	1.2	3.5	6.6	9.5	14.0
Kuwait	0.9	1.5	2.1	3.4	6.4
Libya	2.2	3.0	4.1	5.2	6.5
Nigeria	1.8	2.5	5.1	8.5	12.6
Qatar	0.2	0.3	0.4	0.6	0.9
Saudi Arabia	1.8	3.5	5.7	7.5	17.4
United Arab Emirates	0.9	1.6	2.2	3.9	6.9
Venezuela	2.8	4.7	6.5	9.4	12.0
Merchandise total	20.3	37.0	54.5	89.8	133.2
Services and Private transfers	4.3	4.8	4.5	—	—
Total	24.6	41.8	59.0	—	—

Source: U.S. Treasury Study, "The Absorptive Capacity of the OPEC Countries," September 5, 1975.

Table 11   Payments Balances on Current Account, 1973–75*

	1973	1974	1975†
	(BILLIONS OF U.S. DOLLARS)		
Major oil exporters	6	70	50
Industrial countries	10	−12	1
Non-oil primary producing countries			
More developed	1	−12	−12
Less developed	−9	−28	−35
Total‡	8	19	4

*Goods, services, and private transfers.

†The 1975 projections are subject to considerable uncertainty and should be viewed as rough orders of magnitude.

‡Reflects balances of countries covered here with nonreporting countries, plus (quantitatively more important) statistical errors and asymmetries.

Source: IMF Survey, August 25, 1975, p. 241.

Table 12 Primary Recycling Pattern, 1974 and 1975

	1974	1975 (FIRST HALF AT ANNUAL RATES)
OPEC countries' current balance surplus (accrued)	70	n.a.
Less net charge in receivables from oil companies	13.8	n.a.
OPEC surplus on receipts basis	56.2	33.4
Disposition of surplus:		
Bank deposits	28.5	13.6
"Euromarket"*	22.8	14.0
United States	4.0	−0.2
United Kingdom sterling	1.7	−0.2
Government securities†	9.6	3.6
United States	6.0	2.6
United Kingdom	3.6	1.0
Other investments and loans	14.5	13.2
United States	1.0	2.0
United Kingdom sterling	1.9	0.6
Other countries‡	11.6	10.6
International organizations	3.6	3.0

*Including "foreign currency deposits" in countries other than United Kingdom.
†Denominated "government stocks" and "treasury bills" in source. Does not include any countries other than the United States and the United Kingdom.
‡"Investments, loans to develop countries, bilateral loans."

Source: First two lines, author's estimate based on IMF figures. Rest of table, *Bank of England Quarterly Bulletin*, September 1975, reported in *Economist*, September 20, 1975, p. 84.

INTENSIFICATION OF STAGFLATION

That the price hike coincided with stagflation, that is, worldwide recession and inflation, was almost an accident. OPEC spokesmen maintain that inflation is responsible for driving oil prices upward. The coincidence of timing, whether fortuitous or causally related, was certainly unfortunate. The higher price, entering into costs of producers and consumers, stimulated the cost push of inflation. And the enlarged payments to the oil exporters, of course, diverted consumer expenditures from other products, intensifying the recession. Substantial disagreement exists on potential fiscal and monetary remedies for the "oil drag" or stagflation alone. There is general agreement among economists on the initial price effects, however. To the $102 billion increase (in 1975 dollars) in the price of 27 m.b.d. of oil imports (at $9.71 per barrel) must be added at least an equal amount for the "sympathetic" rise in the price of 29 m.b.d. of non-OPEC oil and the equivalent of 67 m.b.d. in other fuels.[2]

The direct effect of the oil price rise in the United States from 1973:3 to 1974:4 was to raise consumer prices by 3.5 percent.[3] Approximately half of that percentage is probably attributable to the price rise in domestic oil and oil substitutes. The direct effect of 3.5 percent includes the rise of the prices of consumption products using oil and other energy sources but does not include the indirect effects of intensification of the wage-price spiral.[4] The latter effects are more controversial. One estimate stated that with unchanged fiscal and monetary policies, the ultimate effect on consumer prices, direct and indirect combined, would be about 7 percentage points.[5] If domestic oil prices had been uncontrolled, the price effects would have been some 80 percent higher.

While it is very widely believed that an "exogenous" increase of consumer prices, such as the 3.5 percent estimated for 1975:4, does enter into the wage-price spiral, some economists challenge this, arguing that only past wages enter the spiral and not the increases in consumer prices.

When the rise in oil prices, 1973:3 to 1974:4, was introduced into the Federal Reserve Board econometric model, it had the effect of reducing the annual rate of real GNP in 1973 dollars by $40 billion in 1975:4 and by $47 billion in 1977:4.[6] Use of another well-known model, the Michigan model, yielded even higher estimates on the effect of the oil drag—over a $60 billion reduction in the annual GNP rate from 1975:4 on.[7]

If, as was probably true, fiscal and monetary policy became more restrictive in order to counteract the inflationary consequences of the oil price rise, the impact on the GNP would have been even more severe. If the original impact of the 3.5 percent direct effect on the consumer price level were to be gradually squeezed out of the inflationary process, the necessary demand management would have increased unemployment and reduced GNP more.

On the basis of recent studies, it has been estimated that it requires $1\frac{1}{3}$ percentage points of unemployment to reduce the rate of price rise by 1 percentage point. To "squeeze out" the original 3.5 percent direct price-rise effect of increased oil prices would then require about $4\frac{2}{3}$ percentage-point years. That is, a demand-management program that had the effect of keeping unemployment rates 2 percent higher than they would otherwise be, after $2\frac{1}{3}$ years, would lead to consumer prices about 3.5 percent lower than they would otherwise be. By a relationship known as Okun's Law, each 1 percent increase of unemployment is associated with about a 3 percent reduction of GNP produced. So a policy that increased unemployment by 2 percentage points over a $2\frac{1}{3}$-year period would involve a cumulative reduction of GNP by 14 percent of one year's output. For the United States, that would imply a total GNP loss of over $210 billion. That is the equivalent of about 10 years of direct cost of the price increase on United States oil imports (6 m.b.d. at $9.71 for 365 days gives $21.3 billion per year). If, then, our monetary and fiscal policy attempted to squeeze out the oil price rise effects so that at the end of $2\frac{1}{3}$ years the general level of prices was to be no higher than it would otherwise have been, over that period the costs of the demand management in GNP lost through unemployment would be substantially larger than the direct costs of the oil price increase. These computations are controversial, however. Other estimates of the "tradeoff" between the rate of price rise and unemployment indicate even higher costs of a demand-management program. While the exact quantitative relationships remain in doubt, there can be little question concerning two of the principal conclusions. In the first place, there can be little doubt that the oil drag has strongly intensified the worldwide recession of 1974–75. Where demand-management programs have attempted to counteract the upward pressure on the general price level, further intensification has resulted.

The OECD countries have pledged to avoid competitive devaluation and trade restrictions. Nevertheless, many of them have been attempting to reduce their deficits, principally by stimulating exports but also by domestic demand management aimed at reducing imports.[8] France, Italy, Japan, and Britain have all adopted policies that (except temporarily for Italy) do not involve competitive devaluation or trade restriction and yet are designed to eliminate their deficits. After 1975, the current-account deficit of the non-OPEC countries may decline steadily as the OPEC countries spend an increasingly larger proportion of their revenues on imports until perhaps 1979 or 1980, when the deficits might begin to run the other way (see Tables 8 and 9).

REDUCING THE DEFICITS

For the next few years, then, as in 1974, those huge current-balance deficits, dwarfing the dollar deficits of the sixties that brought on the abandonment of the Bretton Woods system, must be accepted and built into the macroeconomic policies of the non-OPEC countries. Yet the policies these countries adopted in 1974 and 1975 did not recognize these deficits consistently. Country after country aimed to eliminate its deficit—by approved means, of course. Many of the leading trading countries of the world did in fact achieve balanced current accounts or even surpluses during the first half of 1975, not solely because of direct policies, but also as a result of their recessions. The deficits were consequently thrown onto the other OECD countries, and, to a larger extent, on the Fourth World non-oil-developing countries (see Table 13).

These other countries, especially the less developed, cannot continue to finance so large a part of the non-OPEC countries' deficit with the OPEC countries. Some relief may be expected as the leading countries enter economic recovery from their recessions and so come to take more imports from the "other" countries. But that would tend to put the leading countries back into deficit, a tendency that is likely to be intensified as the other countries cut down on their imports. In

Table 13   Balance of Payments of Non-Oil-Developing Countries, 1971–75*

	1971	1972	1973	1974†	1975‡
	(BILLIONS OF DOLLARS AT CURRENT PRICES)				
Exports, f.o.b.	38.6	45.6	65.5	91.0	89.5
Imports, f.o.b.	45.5	50.4	69.0	107.5	111.5
Trade balance	−6.9	−4.8	−3.5	−16.5	−22.0
Services, net	−6.3	−7.4	−8.8	−10.5	−11.5
Credits	12.0	13.1	15.7	18.0	20.5
Debits	18.3	20.5	24.5	28.5	32.0
Private transfers, net	0.5	0.9	1.2	1.5	1.5
Balance on goods, services, and private transfers	−12.7	−11.3	−11.1	−25.5	−32.0
Official transfers, net	4.4	5.9	6.1	8.0	9.5
Current balance	−8.3	−5.4	−5.0	−17.5	−22.5
Capital, net	8.5	10.1	11.5	18.5	17.5
Direct investment	3.2	3.0	4.0	3.5	4.5
Aid	3.3	3.1	4.4	4.5	5.5
Euroborrowing§	1.4	2.1	5.6	6.5	5.5
Other¶	0.6	1.9	−2.5	4.0	2.0
Balance on official settlements	0.2	4.7	6.5	1.0	−5.0
Change in reserves	1.2	5.2	7.1	2.5	−2.5
Memorandum items (percent change in volume)					
Exports	—	7.1	14.8	5.0	−4.0
Imports	—	3.6	14.9	13.5	−11.0
Reserves (year-end)	13.4	18.6	25.7	28.5	26.0

*All non-OECD countries except oil-producing countries, the socialist countries of Eastern Europe, the Soviet Union, China, North Korea, North Vietnam, South Africa, Israel, Yugoslavia, Cyprus, and Malta.
†Estimate.
‡Forecast.
§Gross publicized Eurobonds and credits.
¶Including trade credits and errors and omissions.

Source: OECD, *Economic Outlook*, no. 17, July 1975, p. 62.

this atmosphere, the temptation is strong for each country to increase its exports and reduce its imports. But the aggregate non-OPEC current-account deficit will still equal the OPEC surplus; the non-OPEC countries cannot jointly reduce their deficit as long as the OPEC surplus is given. Any attempt would result in a shrinkage of international trade and unacceptably high levels of unemployment.

What steps, then, should be taken? Certainly, a reasonable first step involves the adoption of a consistent set of current-balance targets. *Inconsistent* goals, biased toward low deficits or even toward surpluses, will tend to perpetuate the worldwide recession. But even if consistent targets are adopted in principle, the problem of moving toward them in practice remains. In the current recession, calls have repeatedly been made for some of the leading countries, whose recessions cut deepest, to lead the world out of recession by reflating. At the IMF annual meeting in September 1975, such appeals were directed at West Germany, Japan, France, and the United States.

What is the appropriate current-account target for each country? One approach to this question excludes from each year's current balance for a country the increase in that country's net oil expenditure from some base period, say, 1973. Such adjusted "non-oil" current balances have been computed by OECD for various OECD countries.[9] Such adjustments are clearly too large for our purposes. If all non-OPEC countries aimed at a balance in their non-oil current accounts, they would be aiming at an aggregate net current deficit equal to the *total* cost of the oil price rise, which exceeds the OPEC surplus. These adjustments do not allow for the fact that the OPEC imports have gone up. The target should be equal to each country's share of the total non-OPEC deficit on current account, not its share of the total increase in oil costs.

Suppose, on the other hand, that in 1975 the OPEC surplus is $45 billion and OPEC oil exports are 27 m.b.d. Then the proportional current-balance target for each oil-importing country would be a deficit of $1.67 billion per m.b.d. of oil imported ($45 billion divided by 27 m.b.d.). If, for example, the 1975 United States oil imports averaged 6 m.b.d., the corresponding current-account target would be a $10 billion deficit instead of the $3 billion surplus projected.[10]

While proportional current-account targets are consistent, they suggest many serious questions. If a country, for example, does not receive primary recycling sufficient to finance its proportional target, would it not be reasonable for it to acquire, through its current balance, the shortfall between its primary-recycling receipts and the target? This suggests a different candidate for a current-account target—a primary-recycling target. Such a target can be defined as equal, with opposite sign, to the amount of primary recycling received by the country.[11] Thus, if OPEC countries invest $4 billion of their projected $45 billion surplus in the United States in 1975, then the United States primary current-balance target for the year would be a $4 billion current-account deficit. For countries that receive little or no primary recycling, the difference between the two targets would be substantial.

The basic principle underlying primary-recycling targeting is that any country that achieves a current balance equal to its primary target will then run a current-account surplus with other non-OPEC countries equal to its current-account deficit with OPEC countries minus any recycling received from OPEC nations.[12] That is, its current-account surplus with non-OPEC countries will equal its overall deficit with OPEC countries.

Primary targets will be consistent, however, only if all OPEC-country primary recycling is assigned to a non-OPEC country. OPEC recycling through an extranational entity such as the Euromarket or through international agencies would create problems. However, as long as all OPEC recycling is assigned directly or indirectly to one country or another, the primary targets would be consistent.

Proportional targeting suggests that the United States should aim at a current-balance deficit of $10 billion a year as its 22 percent share of the aggregate non-OPEC deficit of $45 billion a year. A primary-recycling target suggests that the United States should aim at a $4 billion current-account deficit, because it got less than its share ($10 billion) of primary recycling. It should then aim to earn the remainder of its deficit with OPEC through a current-account surplus with other non-OPEC countries, which must have received jointly primary recycling totaling $6 billion in excess of their proportional share. Suppose, on the other hand, the United States should receive a much larger capital inflow from OPEC, say $30 billion. Its primary-recycling target would then be a $30 billion current-account deficit. If its current-account deficit with OPEC was $10 billion, its target current-account deficit with non-OPEC nations would be $20 billion. The implicit assumption in primary-recycling targeting is that secondary recycling should be trade recycling.

Consider a country that receives no primary recycling. It is thereby given a license to do what would be disastrous for all non-OPEC countries—aim at a zero current balance. Clearly, primary targeting ignores the possibilities of secondary capital-recycling.

The leading countries—that is, the most creditworthy—can reasonably be expected to depend more on capital flows, while affording the less creditworthy countries the opportunity to finance the latter's oil deficits partly by capital flows and partly by running current-account surpluses with the former. The ideal targeting arrangement here might begin with primary-recycling targets and then modify the targets to account for the relative creditworthiness of the country concerned. The implicit demand on the leading countries to go into debt (or otherwise transfer claims) in excess of their oil imports appears more reasonable when it is remembered that those countries can, if they wish, adopt domestic monetary and fiscal policies that build up domestic investment to match any buildup of foreign claims. On the other hand, the most disadvantaged countries are neither in the position to increase commercial debt nor to run current-balance surpluses with other non-OPEC countries. For these countries, it is concessionary aid—or the giving up of growth—if they cannot increase their exports to other non-OPEC countries.

What part of any country's oil deficit can most appropriately be paid for by capital transfer and what part by a trade surplus with other non-OPEC countries? The trouble with proportional targeting is that it ignores creditworthiness or other considerations affecting the appropriate balance between trade recycling and capital recycling. The trouble with primary targeting is that it implicitly makes a one-sided judgment—that the primary-recycling decision should determine the capital flows and that trade flows should fill the remaining gaps. In secondary targeting, secondary recycling may be best performed with a combination of capital flows and trade flows.

GENERAL SUMMARY

Global balance requires permitting some non-OPEC countries to improve their current balances at the expense of others better positioned to import capital. The strains on competitive industry would be great but less than if freely floating currencies were permitted. If a rigorously defined floating system were used, it would be equivalent to using secondary targeting with the relative shares of trade recycling and capital recycling determined by the market. Some sensational changes in exchange rates could be expected in order to alter the trade flows to complement the pattern of primary recycling chosen by OPEC and the secondary capital-recycling determined by the holders of financial assets. There probably should be some official international interaction with this process.

The adoption of any consistent basis of targeting shifts the base line from which a competitive depreciation is to be measured. If we define a competitive depreciation as one made by a country in current-account surplus, an endless sequence could ensue of apparently noncompetitive devaluations, as all the non-OPEC countries in current deficit tried to shift the inescapable oil deficits onto each other. However much the non-OPEC countries jointly devalue, they will not eliminate their collective deficit with the OPEC countries. If each non-OPEC country followed the rule of devaluing when it had a deficit on current account (a rule that would be automatic under truly flexible exchange rates with balanced capital accounts), there would be an unending cycle of devaluations as the irreducible deficit with OPEC was passed around among the non-OPEC countries. If, on the other hand, all non-OPEC countries adopted the above rule relative to a consistent current-account target, there would, in general, be an equilibrium set of exchange rates. It is true that this would require each country to arrange for a capital inflow (or a reserve transfer, or some combination thereof) equal to its targeted deficit. That is why secondary capital-recycling is necessary, even if there is a perfect distribution of current-balance deficits according to a consistent specification of the appropriate share of each country in the aggregate deficit with the OPEC countries. Unless consistent targets are adopted, room exists for a good deal of harmful self-deception as to what is and what is not a beggar-thy-neighbor policy.

IV / Financing OPEC Surpluses

The oil-importing countries must pay for their oil imports from the oil-exporting countries, as well as for any non-oil imports, either by export of their own goods and services to the oil exporters or by the transfer to them of claims to assets, real or financial, principally the latter. The claims transferred range over a wide variety of assets. The principal components are present or future claims to money in the form of the customary financial instruments—bank deposits, short-term government paper, accounts receivable, public and private bonds, or other certificates of indebtedness. Claims to real estate and corporate equity have played a small part. Some of these financial claims, when held by the monetary authority of the country concerned, are counted as official reserves, while others, either because of their form or the identity of the holder, are not. For many purposes, the distinction between reserves and other foreign claims held by a country has little relevance: it is the total holdings of foreign claims that count.

The acceptance by the OPEC countries of transfers of claims in settlement of their surpluses on current account is called "primary recycling." As viewed by the oil-importing countries, it represents their being able to pay for part of their imports by a transfer of claims rather than by a direct flow of goods and services.

PRIMARY RECYCLING

The potential problems of primary recycling derive from the possibility that the transfer of claims to the oil exporters in partial payment for their exports may be impeded. This would have serious consequences over the next few years, roughly to 1980, since it is unlikely that the OPEC countries can, until then, raise their joint import expenditures and aid grants to the level of their foreign receipts. If the flow of oil exports is to be maintained at the current real price of oil, primary recycling must take place at a level starting from $70 billion in 1974 to about $45 billion in 1975 but trending downward toward zero by the 1980s, subject to fluctuation in between (see Tables 8 and 9).

The extreme form of a primary-recycling crisis would occur if all oil producers obeyed the call of Algeria, Iraq, and Libya to produce only as much oil as is required to cover their current import requirements.[1] If OPEC producers had obeyed that call in 1975, OPEC oil production would probably be reduced a fraction of current output, with the reduced output selling at a fantastically high price. The recycling problem, then, involves how to prevent this possibility from occurring by ensuring that attractive investment opportunities are available to all producers whose revenues exceed their current needs for making payments.

If the exporters elect to sell less, they will receive more revenues, while the importers will pay more and get less. The non-OPEC countries would find the costs of such a cutback a heavy burden. The FEA estimate of a $33 billion annual loss to the United States for every million daily barrels cut back after the first million may be excessive. However, there can be no doubt that an oil-production cutback of a few million barrels a day would impose heavy costs on the importing countries. OPEC spokesmen have often declared that the erosion of the value of Western currencies, and of investments in those currencies, is likely to discourage the production of oil beyond the amount the oil exporters need to pay for their imports. To wit: "If, in order to meet your needs, you want us to produce more than enough to meet our needs, you should offer us, for our savings, an asset whose value will not melt away. If oil in the ground is worth more than money in the bank, we should leave in the ground the oil we do not need to sell for our current needs for consumption and investment."

Forebodings of crisis have occasionally emanated from the international financial markets whenever it appeared that the swollen flow of OPEC funds would find difficulty in being placed in their customary channels. The full extent of the crisis was seldom played out, but the implication was that further OPEC placements would be turned down by banks that had previously welcomed their deposits. That the consequence might be a cutback in oil production was usually left unsaid, but it was quite definitely implied. It is altogether too easy to imagine an OPEC spokesman declaring: "Today several leading international banks rejected our offers of deposits. If the money you give us for our oil is not worthy of acceptance by your leading financial institutions, we must reconsider the rate at which we are producing oil to exchange for that money." There will undoubtedly be other channels through which the money may advantageously be invested. The solution to the primary-recycling problem is to ensure that this continues to be so.

THE FUNDAMENTAL PROBLEMS OF PRIMARY RECYCLING

A fundamental failure of primary recycling would occur if, because foreign assets were unattractive, the OPEC countries as a group cut back their oil production, if in other words, they preferred oil in the ground to money in a foreign bank. If the level of oil exports does not depend on the attractiveness of foreign assets, no *fundamental* primary-recycling problem exists. Other problems of primary recycling may arise from the imperfect adaptation of the money and capital markets to the primary-recycling process. These may be termed *special* problems of primary recycling, since they relate to specific difficulties in certain sectors of the financial and asset markets, while the fundamental primary-recycling problem relates to the feedback from the financial and asset markets to the level of operation in the oil market. The special problems of primary recycling range over a wide variety of situations. For example, there is said to be a mismatch of lenders' and borrowers' maturity preferences in the various financial markets. Some countries fear the specter of piling debt on debt. Some host countries fear the consequences of the OPEC countries' owning substantial shares of the equity in some important companies or of the real estate in some localities. Some observers discern a special problem arising from the fact that it is not the private citizens but rather the governments of OPEC countries that are destined, apparently, to own substantial—and in many cases controlling—shares in companies operating within the oil-importing countries.

The distinction between the fundamental and the special primary-recycling problems can be most succinctly expressed as follows: Failure to solve the fundamental problem, if it exists, would mean reduced flows of oil and elevated prices. Failure to solve one of the special problems, affecting a particular class of asset in a particular country, would mean that OPEC investable funds would flow to other assets or to other countries.

Even if primary recycling incurred no problems, so that the OPEC countries could easily and smoothly place their surplus earnings in attractive investment outlets, serious problems of secondary recycling might still occur. As illustrated earlier, primary recycling refers to the reflow of funds from OPEC countries to the non-OPEC countries as a group, while secondary recycling refers to the flows of funds among the non-OPEC countries. Secondary recycling is likely to be required because of unbalanced primary recycling, that is, some countries will receive more primary recycling and some less than their net current balance with the OPEC countries. In distributing their funds, the OPEC countries will favor the most creditworthy countries and the Eurocurrency market, which may be thought of, in the present context, as one more country with the special characteristic of recycling out to other countries almost all the funds that flow into it—an almost perfect secondary recycler. The principal problem of secondary recycling is, therefore, how are the countries that receive less primary recycling than their current deficit with OPEC countries to pay for their oil? Aside from drawing down their reserves, these "primary deficit" countries must either obtain more primary recycling, borrow from other non-OPEC countries, or maintain a current-account surplus with them. From the viewpoint of the leading non-OPEC countries, the question becomes: Shall I transfer my good claims to assets to the OPEC countries in exchange for poor claims on the less creditworthy countries, or shall I run a current-account deficit even larger than my share of the aggregate, non-OPEC current-account deficit, so that these other countries, which cannot attract their share of primary recycling, can compensate by running a current-account surplus with me?

The fundamental primary-recycling problem is not likely to arise, only the special problems and those of secondary recycling. The world supply of oil is unlikely to be limited by the unavailability of suitable investment outlets for the producing countries.

Sometimes a similar conclusion is reached on fallacious grounds. It is often asserted that if OPEC runs a surplus on current account, that surplus, and the corresponding deficit of the non-OPEC countries, is necessarily being financed. The assertion is true, but the implication that there cannot be any fundamental primary-recycling problem is false. The accounting identity implies that the non-OPEC countries can run no larger deficit on current account than OPEC is willing to finance by accepting the financial assets the non-OPEC countries can offer. Depending on how much in foreign assets the exporters are willing to accept for their oil exports in excess of what is required to pay for their imports, the amount of primary recycling could be either large or small, with a corresponding variation of the size of the OPEC surplus on current balance. If, as is probably true, the level of OPEC imports does not significantly vary in response to the attractiveness, or lack of it, of foreign investment opportunities, any variation in the amount of foreign assets the OPEC countries are willing to acquire would mean a corresponding variation in the amount of OPEC export receipts. If OPEC members, as a group, should find available foreign assets unattractive and thus keep more of their oil in the ground, they would consequently drive the price of oil higher and higher until the non-OPEC countries' demand for oil became elastic. Fortunately, this possibility is slender indeed.

The main thrust of the OPEC countries' selling strategy involves the *price* of oil. OPEC policies set the government take per barrel for the private companies and a selling price for nationally owned companies, and OPEC members let the marketing capabilities of "their" companies, private and national, determine how much is sold. Almost all the oil-producing countries are likely, therefore, to desire to sell more at the going price rather than less. Their sales being limited by the demand for oil at the price they have set, they are unlikely to further limit those sales because they are displeased with investment prospects in the world financial markets.

Under prospective conditions extending into the 1980s, the producing countries are likely to have excess production capacity well above the volume of their sales at the new high level of prices. Surplus capacity, at its peak in 1974, approached 35 percent of total capacity. Even if there is a strong recovery of sales late in 1975, excess capacity will probably remain at least 25 percent of total developed capacity. And developed capacity at present is far below potential capacity of the fields already discovered. The populous non-surplus oil producers will want to produce more to generate more revenues. The surplus countries, it is true, could make do with smaller revenues but are unlikely to, except to maintain price. There is a substantial probability that within 5 or 10 years the price will not be appreciably higher than the current price. It would seem prudent, therefore, for each country to sell all it can at the current price without cutting the price below the agreed-on level.

Although oil-country spokesmen have declared publicly that the prospective yields on their foreign investments would largely determine whether their countries would or would not increase production and exports beyond their own budgetary needs,[2] their countries' actions suggest a different response. The frantic efforts of Iran to buy developmental projects as fast as possible and the gargantuan Saudi Arabian five-year development project (programmed at $140 billion, though most observers doubt anything like that scale can be attained) seem to imply that in their judgment general industrial, agricultural, and infrastructure productive capacity at home is better than either oil in the ground or money in the bank. If not for the mistakes likely to be committed in so rapid a drive for economic development, it would be hard to fault the wisdom of their choice. Better solutions to the primary-recycling problems, that is, more attractive foreign investment outlets, would, however, permit a more orderly pace in the economic development of the OPEC countries—particularly in Saudi Arabia. Then the countries could acquire domestic production capacity more nearly in line with their absorptive capacity while their foreign investments accumulated earnings abroad. But they appear to fear, and with sound reason if the evidence of the recent past is relevant for the future, that their foreign investments plus accrued earnings will buy less in the future than the planned domestic investments plus their accruals.

When in 1972 Kuwait limited production in order to make its reserves last longer, it may have implicitly adopted a zero discount rate or else developed a sensitivity to considerations other than to maximizing the present value of its assets. Responsible leaders in several oil-exporting countries feel that the explosive expansion of revenues may harm the social, political, and economic fabric of their countries. Political corruption and moral deterioration, they feel, are the inescapable consequences of their countries' sudden enrichment. Their judgment that it would be preferable to go slower remains quite independent of the quality of the available investment assets. Should such judgments prevail, some producing countries might indeed cut production, allowing others a larger slice of the market.

Aside from such noneconomic considerations, cutting

production to maintain price will continue to be the prime factor affecting oil exports. Even if some of the surplus countries, say, Saudi Arabia and Kuwait, should decide either for economic or noneconomic reasons to cut production below what they could sell at the cartel's price, the move would not seriously interrupt the flow of oil. The other producing countries would gladly take up the slack and would be capable of doing so. The principal net effect would be to strengthen OPEC's control over price by easing the pressure of excess capacity. Such a development, of course, would permit the more populous oil-producing nations to raise their incomes faster and thus to proceed more rapidly with economic-development programs.

THE SPECIAL PROBLEMS OF PRIMARY RECYCLING

If the surplus OPEC countries let their surpluses run at a level determined by their sales revenues and their payments requirements on current account, then only the special problems of primary recycling will remain. These may be summed up in a single question: "Where, and in what forms, will they ('the mountain of foreign assets in the hands of the OPEC countries') find their ultimate home?"[3]

In 1974, the where was principally the no-man's-land of the Eurocurrency market, followed by the United States, the United Kingdom, and the rest of the developed world (see Table 12). The United States, with its sophisticated financial and capital markets, was expected to be in a highly favorable position. While it did attract more primary recycling than any other single country in 1974 (but less than the Eurocurrency market), it attracted no more than its share in oil imports (about 20 percent), if neither the accounts payable of the major oil countries nor the deposits in United States–owned Eurobanks are counted as primary recycling in the United States. Its share in 1975 was substantially less than the amount of its OPEC oil imports.

Liquid investments led by a wide margin, followed by grants and loans, with direct and portfolio investments a dis-

tant third (see Table 12). It is expected that portfolio investments and intergovernment loans will, in the future, constitute an increasing share of each year's asset accumulation; the
liquid assets will thus be drawn down as the portfolio investments increase. Plentiful assets are available in the non-OPEC
countries from which OPEC investors can choose. Marketable
financial assets in the OECD countries are estimated at over
$3 trillion, total financial assets over $4 trillion, and physical
assets over $6 trillion. The annual growth of the marketable
financial assets alone exceeds $300 billion. OPEC countries
should be able to acquire some $250 billion of foreign assets
over a five-year period. They have already acquired about
$55 billion in a single year, 1974 ($70 billion counting the
growth in accounts receivable from the companies), and their
cumulated foreign assets are expected to be about $122 bil-
lion at the end of 1975, including $17 billion acquired before
1974.

Special problems of primary recycling are likely to affect
the financial markets but not the supply of oil. These problems involve matching the assets of surplus OPEC countries
with their investment preferences. Banks may find that their
ratios of deposits to capital will not permit them to accept
further deposits from OPEC members, particularly on a
short-term basis. Or perhaps the bankers may not want to
have any unduly large fraction of their liabilities controlled by
a small group of depositors. Host countries may well be concerned over foreign ownership of their business enterprises.
In short, the problem is to adapt the money and capital markets to the primary-recycling process and vice versa. If, for
example, Saudi Arabia insists on short maturities for its
enormous placements, it must expect to drive the short-term
rate down. Readjustment of the maturity preferences of
Saudi Arabia and other OPEC countries was well advanced
before the death in 1974 of Anwar Ali, the principal financial
advisor to King Faisal and the administrator of Saudi foreign
placements. The trend may be expected to resume, and the
share of longer maturity loans and direct investments will grow
in response to ordinary market incentives.

Even though the 1975 rate of accumulation will probably fall well below that of the second half of 1974 (see Table 12), certain financial sectors may be unable to absorb OPEC funds at the same rate as in 1974. On the other hand, the OPEC countries seem to have turned away from certain types of assets in which they invested heavily during 1974. British and United States assets in general seem to have lost favor, while loans to and foreign currency deposits in other countries rose from 35 to 60 percent of the total between February 1974 and January 1975. It was feared, in 1974, that the Eurobanks (particularly those in London) might not be ready to accept as much OPEC money in 1975, principally because they might not be ready to pass on that amount of credit to the potential borrowers. Considerations of maintaining desired ratios of capital to deposits were also expected to enter, but the limited creditworthiness of potential borrowers was expected to be a more serious factor. Perhaps something of that sort occurred, but it looks as though the OPEC countries turned away from the London Eurobanks before the latter turned away the OPEC depositors. In any case, there does not yet seem to be any serious evidence of the OPEG countries' finding a lack of satisfactory outlets for their surplus funds.

The solution to the primary-recycling problem, then, consists in providing financial-instrument vehicles that will satisfy the needs of both investor and host country. Viewed from OPEC's side, the problem is the standard portfolio problem—into what assets shall I put my wealth? The OPEC investors seek, like all international investors, safeguards against four main types of risk:

1. Loss of liquidity
2. Exchange-rate risk
3. Inflationary erosion of real values
4. Default or confiscation risk

It is appropriate to consider what new financial arrangements might improve the current situation.

NEW FINANCIAL ARRANGEMENTS

The whole international financial system would undoubtedly work better if international obligations could be put into a value-preserving form, such as special drawing rights (SDRs) indexed to maintain purchasing power.

Strong arguments exist for the development of institutional arrangements to protect OPEC lenders, or anyone else, against exchange-rate risk. OPEC members may be prepared to accept interest-rate differentials or asset-appreciation prospects as premiums for bearing such risks. Even much more important than protection against exchange-rate risks is protection against inflation risks. Borrowers and lenders should have the choice of dealing either in ordinary money or indexed money, with different interest rates. (If provided only for OPEC creditors, however, the indexed money would have the disadvantage of being subject to default.)

The most attractive form of indexed deposit might be with the International Monetary Fund (IMF) or the newly constituted International Central Bank, in SDRs indexed by a suitable price index—probably of internationally traded commodities.[4] Let us, for convenience, call such a reserves unit the constant-valued reserve unit (CRU). The IMF (or the International Central Bank) would maintain an exchange rate between the SDR and the CRU based on an appropriate index of internationally traded goods. Monetary authorities could freely move their reserves on the books of the IMF between CRUs and SDRs at the current rate of exchange. Possibly similar privileges might exist on the debt side, within limits set by the IMF through international agreement. Just how the interest rate differential between CRUs and SDRs would be determined is a question of some complexity into which we need not go here. For illustrative purposes, however, perhaps both CRU and SDR interest rates might be set by market forces against the background of the total volume of SDRs and CRUs together being set by the IMF on the basis of world monetary and trade conditions. The essential feature of such an arrangement is that central bank or other monetary authorities have the power to balance their reserve portfolios between high nominal rates and lower real rates of interest, with corresponding shifts in the sensitivity of their portfolios to inflationary risk. OPEC countries could then invest their foreign assets in inflation-proof form up to the full amount at which they are willing to accept the differential rate of return between a nominal and a real rate of interest.

If they took advantage of this opportunity to a major degree, each of the non-OPEC countries could then determine how its monetary authority would handle its end of the transaction. There would be complete buffering between the domestic financial markets in the OPEC and non-OPEC countries, respectively. It is presumed that under these arrangements the SDR interest rates would come into balance with the domestic nominal interest rates in the various countries and actual or potential forward market rates in the various currencies. While the full design of such a system is outside the main focus of this study, the desirability, in principle, of such a system is worth noting. For the primary-recycling problems in particular, and possibly for the international monetary system in general, it would be most useful to have an indexed reserve asset.

Such an asset would go far toward meeting the OPEC countries' concern for protection against risk of default or confiscation as well as inflationary erosion of value. Under extreme circumstances, such as war, almost any foreign asset is subject to sequestration or "freezing," certainly in enemy countries and possibly even in friendly or allied countries. Even in wartime, however, international monetary reserves may be respected. Short of war, there are few circumstances in which it is likely that the claims of a country embodied in its reserve at the IMF would be defaulted. On the other hand, if all the members of the IMF were to have the right to borrow SDRs and CRUs up to some ratio of their quotas, there would be a risk of default, and there is some concern as to how that risk would be distributed. If it were set in proportion to current quotas, the United States might reasonably object. If it were set in proportion to the holding of SDRs and CRUs, their attractiveness would be limited. We are now encountering, perhaps prematurely, the fundamental problems of secondary recycling: how to minimize the risk of default and how to distribute that risk. The antipathy, or at least the lack of enthusiasm, of the United States toward the IMF special facility can probably be attributed, at least in part, to the fact that since the OPEC lenders to that fund are protected by the IMF as a whole, the United States bears a large share of the contingent liability.

Theoretically, the solution of the primary-recycling problem through the facilities of the IMF has many attractive features. Practically, the proposal has many difficulties. The advantage of further internationalizing relationships are substantial. The risk of the less-creditworthy countries defaulting can be tempered to a certain extent by their working with the IMF to maintain their solvency. But the IMF is unwieldy in operation and nearly impossible to reform. Whether it could adopt to so radical a change as an indexed SDR is questionable, to say the least.

Other new institutions have been advocated to help with primary recycling and possibly to reduce the need for secondary recycling. An "OPEC Fund for Government Securities" and an "OPEC Mutual Investment Trust" have been proposed. Institutions such as these have much to recommend them. They reduce the risk of default to the investor and the risk to the investee of undue concentration of ownership. Risk of inflation can be offset, possibly, by the returns earned, exchange risks possibly, but not entirely, by the nature of the assets. These proposals seem to have had limited appeal to the OPEC countries, however, being reportedly reminiscent of a colonialist past.

For the oil-importing countries, the advantage or disadvantage of OPEC investments varies with the country. For most, foreign investment in general, and OPEC investment in particular, is welcome.

For each country by itself, an additional dollar of foreign investment can enable it to consume more, invest more at home or abroad, increase government purchases of goods and services, and build up reserves, in any combination aggregating one dollar. It can achieve these ends by various combinations of fiscal, monetary, and foreign exchange policy. Countries, such as Switzerland and Germany in 1974, who did not want to do any of the things made possible by an inflow of funds, may indeed want to discourage such inflows of funds. But most of the oil-importing countries are in the opposite situation and so will find an inflow of OPEC investment funds advantageous.

For those countries that are net importers of OPEC capital, it would be best, it has been argued, to adopt policies consistent with full employment, leading to increased domestic investment roughly equal to the inflow of OPEC funds. Such a policy can be supported by the following considerations. The continuing inability, until about 1980, of the OPEC countries to accept imports up to the full value of their exports at the new high prices means that the the oil importers must run a current-account deficit. Those oil-importing countries to accept imports up to the full value of their exports at the new high prices means that the oil importers their domestic investments correspondingly, thus building up additional domestic assets to match their increased external liabilities. If they do so, however, almost the full weight of the oil price rise will fall upon current consumption. That part of the increased costs of oil that goes into enlarged exports to OPEC comes primarily out of consumption. Alternatively, an oil-importing country might use the inflow of OPEC funds to maintain consumption closer to its pre-price-rise relation to the national income, cushioning the effect on consumption of the original deterioration brought on by the oil price rise. But this move would throw a greater burden on future consumption or investment when the OPEC claims were liquidated.

The host country usually captures through taxation a large fraction of an investment's economic contribution to the national output. Every dollar of annual income earned by a foreign investor in the United States is matched by an annual contribution from his investment to tax revenues—national, state, and local—that amounts to more than a dollar. There is nothing special about *foreign* investment in this respect. Every dollar of income to the owner of a domestically owned investment is also matched by more than a dollar contribution to tax revenues. In evaluating the tax benefit of foreign investments made in a country, allowance must be made for the tendency, if any, of such investments to displace domestic investment. If, on the other hand, a dollar of capital inflow can support an additional dollar of investment in the country, it will make an important contribution to the country's subsequent national income.

If OPEC investors wanted to invest in the United States and in most other oil-importing countries, they would be welcomed, provided the conditions of their investment were not identifiably hostile to the national interest of the host countries. For the most part, the OPEC investors will find it to their own interests to spread their investments widely through the financial markets consonant with the national interest of the host country.

THE ARAB BOYCOTT

The Arab boycott against Israel and against firms believed to be sympathetic with Israel, possibly because they have Jewish officers, poses a difficult, complex problem. Even more subtle are the effects on firms in non-Arab countries who, desiring to do business with the Arabs, take care to preserve the ethnic acceptability of their staffs. Some firms may use the presumed reluctance of Arab investors to deal with Jewish firms as a competitive weapon, even though the Arabs may not raise the issue.

The financial markets are sufficiently well organized that the traditional penalty against discrimination can be expected to evolve from the operation of the markets themselves. If, for example, the Arab investors were to favor the stocks and bonds of certain companies for portfolio investment as opposed to others, so that their investments would tend to drive up the price of the stocks and bonds of their favored companies, the rest of the market would soon find that it paid to invest in these other companies.

The impact on commercial and industrial firms is likely to be more serious, however. Arab purchases represent lucrative sales opportunities, and there are already signs of forces operating powerfully, if subtly, in certain export markets. Where there is keen rivalry for sales, no firm likes to be disadvantaged. Stronger measures are probably needed to ensure against discrimination contrary to our national policy.

DIRECT INVESTMENTS

In regard to a wide variety of asset categories in which OPEC investors may continue to invest, there should be no serious problems affecting the public interest. Bank deposits, government securities, and portfolio purchases of equities fall in this class. Only one type of investment can be expected to raise serious problems in the host countries—direct investment. Direct, as distinct from portfolio, investment involves a significant degree of control. Acquisition of equity ownership in real estate is direct investment, as is purchase of over 10 percent of the voting stock of a corporation.

Direct investments have so far accounted for only a tiny fraction of the OPEC countries' acquisitions in the oil-importing countries. But they have accounted, unfortunately, for a major share of the public reaction. A quarter interest in the steelmaking division of the Krupp interests for about $100 million, 14 percent of Daimler-Benz for about $400 million, a controlling interest in Detroit's Bank of the Commonwealth, rumored at only $10 million, a projected but long-postponed $300 million investment in Pan American Airlines (Pan Am), and a $17.3 million investment in an island off the South Carolina coast have made headlines. These direct investments are far from characteristic of the bulk of OPEC investments, of which they represent a tiny part. Of the $11 billion total OPEC investment in the United States in 1974, about 1 percent, or $100 million, was direct investment, principally in banks and real estate.[5] Nor is there any compelling reason why the OPEC countries should make large direct investments in American corporations. Direct foreign investment requires a special sort of motivation—usually a unique competence on the part of the purchaser in conducting some line of business that can be extended to the host country. Or the purchased company may have some special competence that the purchaser can use in the home country. Such situations are not likely to arise between the OPEC countries and the leading countries of the West. The OPEC countries are likely to be interested in corporate securities, principally for portfolio purposes. Even if United States regulations should discourage OPEC direct corporate investment entirely, there would be no great loss either to the United States or to OPEC.

While little OPEC investment would be lost by placing restrictions on direct foreign investments, such restrictions are undesirable. They are not necessary, and their side effects are harmful. They would necessarily be imposed as general restrictions on all foreign investors rather than specifically on OPEC investors. They would undoubtedly stimulate the general imposition of reciprocal restrictions on American direct corporate investments abroad. A large volume of American direct corporate investment abroad would be harmed by the general adoption of domestic restrictions on foreign direct investment. Furthermore, we have treaty obligations with many nations, and they with us, requiring that foreign investors be given the same treatment as domestic investors, with certain obvious exceptions.

The Ford administration claims informal understandings with the principal OPEC countries, whereby they consult the United States government before making any important investments, a most suitable way to handle such problems.[6] Iran, for example, sought advanced concurrence of the United States government in its negotiations with Pan Am. Takeovers of control over large corporations are like a city built on a hill: they cannot be hidden. The attendant publicity would, in most cases, be highly unwelcome to OPEC investors. On the other hand, extensive portfolio investments can be acquired quietly, at least in the United States. Because of the public nature of the takeover attempts, there need be little fear that they will escape governmental monitoring. Informal consultation by the United States government with potential buyers into control of American companies is likely to be generally effective, especially when the potential buyers are foreign governmental agencies.

The OPEC countries likely to be acquiring the most important volume of assets in the United States will generally operate as foreign governmental buyers. The difference between a purchaser who is a private entity, individual or corporate, and a purchaser who is a foreign government is a significant one. The reason why we need be little concerned in general by private foreign takeovers of American corporations is that the corporations' activities are still completely subject to American law and circumscribed by the operation of market forces. But the situation is more delicate if the controlling stockholder is a foreign government. This problem is recognized in a minor way by the exemption of foreign dividend payments to foreign governments from the withholding tax levied on dividends paid to private foreign stockholders. While it is not a major policy concern, it would still seem best, on the whole, not to have foreign governments in control of American companies. We are undoubtedly ready to respect the right of other nations to reciprocate by informally (or perhaps even formally) excluding the United States government from owning their domestic companies.

The underlying justification for according foreign ownership the same rights as domestic is not only reciprocity. More important is the expectation that the company will be operated for economic ends. When the ownership and control rest in a foreign government, there is some presumption that political ends will enter into the operating considerations of the enterprise. In fact, even for portfolio investment, if the investor is a foreign government, it would be more congenial for the host country if the ownership of equity is indirect, permitting some form of intermediation through host-country *nationals*. For then the underlying expectation that the investment is for profit and security of capital only and not for ulterior motives is most likely to be valid.

Financial spokesmen for the OPEC countries complain that they do not have any clear information on what is acceptable and what is not acceptable in the capital markets of the

West. Yet the United States government has already issued clear guidelines for foreign investors. The OPEC spokesmen do not, evidently, refer to these but rather to the prospect that new restrictions may be imposed in response to the adverse publicity given to OPEC acquisitions. There is so much excitement in the public press at their every move, they say, that they wonder what they can do. They say they would prefer well-defined restrictions to the boundless threat of undefined restrictions to come. It is clear that if they want to attempt a takeover of, say, General Motors or IBM there would be an uproar. But they should certainly be assured that if they want to acquire diversified portfolios of corporate stocks and bonds, as well as government securities in various Western countries, they are welcome. They know this already, but it would help to make it clear publicly.

SUMMARY

The special problems of primary recycling are principally problems of market adjustment. There is a broad range of foreign assets that OPEC countries may safely and profitably acquire with their surplus funds without endangering the sovereignty of the host country. Special problems arise principally with respect to the possibility that noneconomic objectives may be introduced in financial affairs and in direct corporate control. Special informal and, in some cases, formal measures may be necessary to forestall these problems. Such measures need not impair the welcome extended to the OPEC countries to acquire satisfactory assets in the United States and in other countries. The surplus OPEC countries will then be able to place their surpluses wherever, in their judgment, the value of those funds can be best maintained or increased, pending the day when they can be used better at home to support investment or consumption or government activity.

V/Paying for Oil

If there were only one non-OPEC country, secondary recycling would not be needed. Or if each oil-importing country received its proportional share of primary recycling, it would face no special problems other than the current real burden of its increased exports and the eventual real burdens of debt servicing and repayment. But most countries have been receiving less in capital from OPEC members than the amount of their oil-import deficits with them. The gap is explained by the large OPEC deposits in the Eurocurrency market, in the United States and the United Kingdom, and in trade-receivable credits from the oil companies ($46 billion of $70 billion in 1974). An oil-importing country's net current balance with the OPEC countries may be called its "primary balance." A non-OPEC country's primary balance is identical with its overall balance with the OPEC countries as a group, that is, the sum of its balances on current and capital accounts.

Most non-OPEC countries run a primary deficit, which can be sustained by a combination of three strategies:

1. Secondary trade recycling, that is, running a current-account surplus with other non-OPEC countries
2. Secondary capital recycling—receiving a net capital inflow from other non-OPEC countries
3. Or, as a final resort, drawing down reserves

The primary deficit non-OPEC countries can run a current-account surplus with the primary surplus non-OPEC countries only to the extent that the latter run a corresponding deficit with the former. The capital inflows into the primary deficit countries from the surplus countries must match—indeed, must be precisely the same as—the capital outflows from the surplus countries to the deficit countries. (The Euromarket may be considered a primary surplus country.) The problems of secondary recycling arise from the unwillingness or inability of the pertinent nations to adopt consistently complementary strategies. A primary deficit country may eventually find that to maintain its economic viability it must either draw down its reserves or cut back its imports. The import cutback need not be, and almost certainly would not be, confined to oil. Given the importance of oil in maintaining a broad range of activities, import cutbacks are likely to be directed mostly to other commodities.

The success of any form of secondary recycling hinges on recipients of primary recycling, who receive more primary capital flows than their current deficits with OPEC (primary surplus countries), transmitting that excess to other non-OPEC countries either through trade recycling or capital recycling. Both modes of secondary recycling have their drawbacks. Primary surplus countries generally are unwilling to sacrifice overseas or domestic markets to other non-OPEC countries in order to permit the latter to run a current-balance surplus with them. The alternative of capital flows may also be unattractive in certain cases because of the lack of creditworthiness of the prospective borrowing countries.

The secondary capital-recycling problem can be reduced to a single word: creditworthiness. And that, in turn, can be reduced in all but a few countries of the Fourth World to yet another word: liquidity.

Within a country, a firm is solvent if its assets exceed liabilities. But countries whose assets exceed their international liabilities, by even a very large margin, may nevertheless not be creditworthy because they may become unable to honor foreign commitments. If claims to domestic assets in each of the countries concerned could be truly internationalized, there would be no problem of creditworthiness and no secondary capital-recycling problem. When, then, should it not be possible for foreign holders of stocks and bonds, or indeed real property, in Britain or Italy or Greece or Bangladesh to sell these assets to buyers outside the country and thus provide the foreign exchange to finance a deficit on current account?

Some observers maintain that, except possibly for the Fourth World, private financial markets will be able to handle adequately the secondary recycling required. And to date, that has been the case. Indeed, most non-OPEC countries (including, for obvious reasons, the reserve countries) have not had to draw on their monetary reserves to meet their enlarged current deficits (see Table 14). To a certain extent, this fact is a misleading consequence of balance-of-payment definitions. Foreign borrowings by governments and by public corporations have, in many primary deficit countries, contributed substantially to financing current balances. Such borrowing does not, as it well might under other definitions, count as a drawdown of reserves.

Meanwhile, there has been a huge increase in the reserves of the OPEC countries as a group in the form of

Table 14   Distribution of Monetary Reserves 1950–75*

	1950	1960	1970	1973	1974	1975
			(BILLIONS OF SDRs)			
Industrial countries	36.8	48.5	65.8	96.0	98.1	100.2
Primary producing countries						
More developed	3.7	3.7	8.7	20.1	17.4	16.4
Less developed, non-oil-exporting	8.5	7.2	12.9	23.3	24.6	24.4
Major oil-exporting	1.3	2.4	5.1	12.0	38.1	41.2
Total	50.3	61.8	92.5	151.4	178.2	182.3
Memo						
Exchange rate—United States Dollar per SDR	1.00	1.00	1.00	1.21	1.22	1.25

*At year-end, except 1975 (March).

Source: IMF, *Annual Report*, 1975, p. 36

foreign-exchange holdings. The net effect therefore has been a large increase of world reserves generated principally in the Euromarkets and held by OPEC countries. The assets behind these increased reserves are the increased debts of the non-OPEC countries, largely to the Eurobanks but also to domestic banks in reserve centers. The one-sided nomenclature system for reserves largely explains the huge growth of reserves: the OPEC short-term holdings of foreign currencies are counted as reserves, while the debts of non-OPEC public and private borrowers are not subtracted from the reserves of the non-OPEC countries.

The limits of creditworthiness are set by the prospective inability of a country to bring its domestic absorption (consumption plus investment) into proper relation to its national income. It is easy to project the way this can be done. Such scenarios are the staple fare of development studies. An attainable growth rate of output can be projected, with a certain proportion of the increased GNP assigned to domestic absorption and the remainder to reduction of the foreign deficit on current balance. Therefore, growing surpluses on current foreign balance which can be used to whittle down the accumulated debt can eventually be achieved.[1]

The Achilles' heel in this case is the assumption that as growth proceeds domestic absorption grows less rapidly. This in fact may be possible in some countries but not in others. The problem of ensuring that a substantial part of any growth of output goes to reduce the deficit on foreign balance is of a piece with the problem of demand-pull inflation, and it may have affinities with the problem of cost-push inflation as well. Some think it can be handled simply by exchange-rate policy.

During 1974, it was feared that many oil-importing countries would face difficulties in financing the increased costs of their oil imports. Some observers questioned the capacity of private money and capital markets to handle the required capital flows. Those who favored government participation did so for two reasons. The first concerned the possibility of special problems in primary recycling, specifically that the limits on the amount of intermediation the Euromarket and the domestic banking systems could perform between short-term OPEC depositors and long-term non-OPEC borrowers would be reached. That would create both primary- and secondary-recycling problems. The second, and more serious reason, was the possibility that certain countries and their private borrowers would become uncreditworthy in the international financial markets. That would make secondary capital-recycling to those countries difficult or impossible. Such countries would then be forced into trade restrictions and competitive devaluations, passing the deficits on to other countries, which in turn might be pushed into a similar pattern of behavior. A breakdown of international financial and trade structures could ultimately be the result of this process.

Numerous suggestions to forestall such a crisis were offered. It was proposed, for example, that the OPEC countries should extend credit to each oil-purchasing country in the same proportion to its purchases as the ratio of the OPEC current-balance surplus was to total OPEC oil revenues. This would generate balanced primary recycling, so that secondary recycling would be unnecessary. Such suggestions did not appeal to the OPEC countries, which would have had to bear the credit risk.

The Kissinger-Simon safety net, though it has not yet become operative (there are some who doubt that it ever will), appeals to some observers.

For fully a year after the onset of the oil price crisis, the United States was opposed to secondary-recycling proposals. Its declared objective was to get the oil price down, and it was apparently feared that arrangements to facilitate secondary recycling would weaken support for measures to build up consumer-nation solidarity in getting the price down. In October 1974, there was a complete tactical about-face, though still directed toward the same strategic end—getting the price down through consumer-nation solidarity.

In a pair of widely heralded speeches in November 1974, Secretaries Kissinger and Simon proposed a secondary-recycling scheme, the Financial Support Fund (FSF), that came to be called the safety net. The FSF, open only to members of OECD, was scheduled to come into operation late in 1975 upon ratification by countries holding 90 percent of the quotas, which total SDR 20 billion (approximately $25 billion). Italy's proposed quota would be, for example, SDR 1.4 billion; Denmark's, SDR 240 million. Any member's borrowing up to its quota would require a two-thirds majority vote (by quota); up to twice its quota, a 90 percent majority; and over twice its quota, unanimity. The borrowing country would be required to demonstrate that it was in financial difficulties, had made the fullest appropriate use of other resources, and its economic policies were consistent with the objectives of the FSF and included measures needed to redress its external financial situation. United States government spokesmen advocating support of the FSF stress that the borrowers would have to show that they were making a strong effort to conserve energy and develop new energy sources.[2] This condition is included in the objectives of the Facility but not explicitly in the publicly announced terms for borrowing.[3]

Less ambitious programs have also been launched with varying degrees of success. A new Oil Facility was established in June 1974 at the IMF, which was empowered to borrow up to SDR 3 billion (roughly equal to $3.6 billion) from member countries in a position to lend and to relend the money to countries who required assistance in meeting their growing oil bills. Most but by no means all of the money was supplied by the surplus OPEC countries. The lenders received and the borrowers paid 7 percent on their loans, a rate of return that some viewed as rather high, since the loans were backed by the entire credit of the IMF. On the other hand, in the first year of operation worldwide inflation was high enough (12 percent on consumer prices in the OECD countries) to make the real rate of return (the nominal interest rate minus the rate of price increase) negative.

Small as were the resources at the disposal of the IMF Facility relative to the total recycling requirements, they were more than enough to meet the demands of those countries qualified to borrow under its terms, and SDR 400 million of the 1974 allocation were carried over into 1975. In addition, the IMF voted in 1975 to borrow an additional SDR 5 billion for loans by the Facility. There seems, however, to have been at least some initial reluctance on the part of potential lenders, and as of September 1975, the SDR 5 billion loan target had not been fully subscribed by potential lenders.[4]

Meanwhile, a United States suggestion had been accepted in principle at the IMF for a trust fund to provide financing for the neediest countries from the profits to be realized by selling one-sixth of the IMF's gold at market prices.

Other international agencies also participated in secondary recycling. The World Bank borrowed $565 million from OPEC countries in its fiscal year 1974 and almost $2 billion in fiscal 1975 to be loaned out to the developing countries. In October 1974, the EEC announced a fund of $3 billion to be borrowed from OPEC countries and lent out to the needy among its members. It ran into difficulties, however, as did a similar UN program. The latter program attracted loans of only about $275 million, a serious blow to its ambitious plans.[5]

Secondary-recycling arrangements like the Kissinger-Simon Fund pose the problem known in the insurance business as "moral hazard." If a country is to be protected from the adverse effects of its improvidence, what incentive will it have to mend its ways in the future? In one form or another, solutions to the secondary-recycling problem require the economically strong to support the economically weak, and there is a reward for helplessness. Secondary recycling to those countries that should be doing something about their deficits, it is argued, relieves them of the necessity to take the steps they should take.[6] This consideration may possibly undermine political support for the Kissinger-Simon financial solidarity safety net in the United States itself. Those who would like the United States to bear less of the risk suggest recycling schemes in which the OPEC countries would take more of the risk. But it is not clear how they are to be induced to do this. Clearly, it would be best if, as previously mentioned, claims to domestic assets could be internationalized, so that sound domestic assets could be transformed into sound foreign assets through private capital markets. The question is what institutional changes could bring about such a dramatic improvement of the international capital markets? Failing that, or if it is achievable only to a limited degree, governmentally sponsored institutions such as the safety net may be necessary to close the gap and recycle to the less creditworthy countries the capital flows necessary to support an appropriate current-balance deficit. How large a deficit is appropriate for each non-OPEC country is a complex question, but it is virtually certain that those countries that have the biggest current deficits relative to their exports should have smaller deficits—as they would if those countries that have the biggest surpluses ran selected deficits instead, in some consistent overall pattern.

This very problem inspired the creation of the International Monetary Fund. But the IMF was designed with transient debt conceived in terms of a year or so, while in the current context, it must cover transient debt from five to ten years. Furthermore, the centerpiece of the Bretton Woods arrangement was fixed exchange rates, which have now been abandoned for floating rates. But if rates were really allowed to float in the current context, there would, in the absence of huge capital movements, be disastrous results. No reduction of exchange rates in the currencies of the oil-importing countries relative to those of the OPEC countries could be expected to bring the OPEC current account to a zero net balance. Capital movements, rather than floating rates, must be depended on as a last resort in the current attempt to balance accounts.

The secondary-recycling problem will come proportionately closer to resolution as the international capital market approaches the status of a perfect market. In such a market, ownership of an asset in any country could be transferred to any other country. If such transfers were easy, secondary capital-recycling could be achieved through them—there would be no secondary capital-recycling problem.

REFORM OF THE INTERNATIONAL FINANCIAL SYSTEM

The first steps involve strengthening the intermediation structure. If the international banking system, or the national systems for that matter, were afforded adequate international "lender-of-last-resort" support, intermediation could be obtained via those systems. An asset that is secure transnationally is most assuredly needed. International branch banking or branch investment houses, together with appropriate national legal structures, could meet this problem if there were an international lender of last resort. And such arrangements need not involve OPEC. A United States bank would, for example, be quite willing to advance money, at favorable terms, to its Italian branch if it knew that the branch's liquid Italian funds could be converted to dollars on demand. The Italian branch could invest as safely and profitably in Italian assets as its skill permitted. With its liquidity assured by a lender of last resort and with exchange rates no more unstable than could be covered by forward transactions, a basis would emerge for capital movements into Italy. The critical risk, as things are now, lies in the possibility of adverse regulations that might trap assets in a form where they are unduly subject to exchange risks. This simple example is offered to show that the secondary-recycling problem arises principally from the imperfections of the international financial system rather than from the lack of inherent creditworthiness in the relationship of a country's assets to its foreign liabilities.

Farmanfarmaian, among others, suggests creation of an OPEC-financed fund for government securities and a mutual investment trust. Why need these institutions involve OPEC? If there were investment opportunities for such institutions, countries might seek their financing through the general financial markets and let OPEC funds go where they want— perhaps directly to these institutions or perhaps to banks or other investment trusts. From the point of view of secondary recycling, a set of institutions is needed that can permit foreigners to invest safely in Italy or Denmark or even India or Bangladesh, so that German or American or British or Euromarket capital might be invested in those countries, while the OPEC countries invested in Germany, the United States, Britain, or the Euromarket. If the OECD countries, in particular, had such capital markets, there would be no problem among them of secondary capital-recycling. The ordinary functioning of a market would redistribute any imbalance of primary OPEC investments, achieving balanced recycling of both primary and secondary together. The problem is to define what must be done to get international capital markets to operate in this way.

REFORM OF THE INTERNATIONAL MONETARY SYSTEM

The breakdown of the international monetary system in August of 1971 and thereafter has been grossly exaggerated. The spirit of Bretton Woods lives on, even though one of its bad habits—fixed parities subject to occasional fracture— seems to have been abandoned for good, and for the better. Another of its habits often judged to be bad—its addiction to the dollar as an intervention currency, hence as a reserve asset—still persists, though the intention has been declared of abandoning it. A basis for terminating the dollar habit has already been laid in the SDR, but it remains to be seen whether the necessary additional steps will be taken to displace the dollar by the SDR or even the CRU suggested earlier.

Within the Bretton Woods framework, international trade and capital movements attained a scale and freedom greater than the world had known. The system, in its "broken-down state," has so far handled the massive payment imbalances generated by the oil price rise well. Floating exchange rates, countenanced as a temporary expedient (the IMF rules do not allow for approving floating), seem here to stay—it is managed floating, but floating nonetheless.

The oil-payments emergency has indeed provided a respectable cue for the Committee of Twenty, which had been charged with the task of reforming the international monetary system, to depart the stage with its work undone. It wasn't the Committee's fault; 1975 is not 1944. Then, the United States could indulge its penchant for written constitutions at Bretton Woods as well as at San Francisco a year later. Whether Lord Keynes had a better plan than Harry White is still a question worthy of debate. But White had the dollars to offer, and Keynes had dollars to get, so the debate was not decided on the merits of the argument. Today, the United States does not have the exclusive command over potential reserves that would justify (with the absolute logic that control of the purse inspires) its writing the rules on its own.[7] So the evolution of the international monetary system is free to proceed without an explicitly reformulated constitution. The system evolves and survives; its adaptations are breakdowns only of outworn features.

The spirit of Bretton Woods is embodied in a system of international cooperation in monetary affairs that can support flourishing world trade and free international capital movements. Whether those ends can best be realized through fixed exchange rates, floating rates, or, as seems to be the case, through rates that are half-fixed and half-floating is a technical question. Indeed, managed floating rates would seem to be as satisfactory a synthesis of the thesis of stable rates with the antithesis of adjustable rates as any of the unspecified alternatives the Committee of Twenty may have had in mind in calling for rates that are "stable but adjustable."[8] This study can hardly undertake to carry on the job the Committee of Twenty laid down on June 14, 1974. But it may well concern itself with the general lines of development in the international financial system that could best meet the problems laid bare by the oil price rise.

The most valuable contribution that the IMF or some other international agency could make to the problem before us is the implementation of measures to prevent the intensification of recession through beggar-thy-neighbor policies, adopted to deal with the widespread incidence of deficits on current balance caused by the oil price rise. We have already considered how a consistent set of current-balance targets could be constructed. But we left unanswered the question of how to ensure that consistent and productive policies would be followed by all the countries concerned. The IMF might reasonably be charged with the procedural responsibility for working toward that end. The IMF has already adopted the suggestion of the Committee of Twenty that it invite its members to "take the pledge" by subscribing to a Declaration on Trade Measures. Under that declaration, a member represents that, in addition to its preexisting obligations under the Articles of the Fund, "it will not on its own discretionary authority introduce or intensify trade or other current account measures for balance-of-payments purposes that are subject to the jurisdiction of GATT. . . . without a prior finding by the Fund that there is balance-of-payments justification."[9] Clearly, that pledge does not go far enough. It leaves implicit the notion that balance-of-payments measures other than trade restrictions (and in other connections, competitive devaluations) are to be countenanced, while we have seen that *any* measure that reduces a non-OPEC country's deficit on current balance will be taken at the expense of the current balance of another non-OPEC country, given the OPEC current-balance surplus.

One important function of the IMF has been Fund staff consultations with member countries concerning measures to deal with foreign balance problems. It is increasingly recognized that foreign balance problems cannot be dealt with separately from the monetary and fiscal policies of the country concerned. Too often this recognition has stopped at the verbal level. Clearly, what is required in the current context is a sort of wholesale consultation with the IMF membership on the coordination of foreign balance policies and of general macroeconomic policies. Whether this might best be done by a joint conference or, preferably, by a coordinated system of bilateral meetings between the Fund staff and each country concerned is a decision to be made. A general Fund meeting on the substantive aspects of the problem would be the best way to kill the proposal. On the other hand, under present conditions, consultations with members are too likely to stop short of those measures that in the end will determine whether recessions will deepen into depressions or whether recoveries will be speedy or slow.

At best, such consultations must be far from satisfactory. Just where any country should draw the line between its anti-recession and its anti-inflation measures, no one can say with authority, though many are quite willing to try. But it is to the interest of all that in this process there be a voice to speak for the interests of the world at large. While the IMF, or any other international agency—say the OECD—is unlikely to be given any real authority over national policies, there will be a certain influence exerted by the process of requiring the domestic economic authorities to explain to the representatives of the international community that their programs and targets are consistent with the interests of all. The exertion of explicit pressures by the international agency would seem to be out of the question except against practices foresworn in the Articles of the Fund or through GATT, but there is still some influence to be exerted by an appeal to common action in the common interest. There is a real job to be done, and if a better way of achieving consistent and constructive behavior exists, it should be practiced. Meanwhile, it would be much better to have such consultations vigorously pursued than not, even though the only sure, realistic sanction is a report to the world on the progress of consultations.

The need for better international coordination of national economic policies and, indeed, the desirability of following *international* economic policies is being increasingly recognized. At the time of this writing, an international economic summit, a meeting of the heads of state of the seven leading non-OPEC countries, is being planned. Certainly, the most valuable outcome of such a meeting would be a mechanism for international coordination of fiscal and monetary policies in general and balance-of-payments policies in particular.

COUNTERBALANCING SHIFTING CARGO

Many appraisals of the petrodollar problem have pointed to a collapse of the international financial system as a possible consequence of the oil price rise. These appraisals emphasize the failure of recycling; the remedy should be to make recycling work rather than to alter monetary institutions.

Some observers have, however, expressed concern over the "shifting-cargo" problem with respect to OPEC short-term balances. This refers to two possible dangers. The first is the threat to the banking systems, national and international, of a sudden massive demand for cash by OPEC depositors. In principle, this problem can be met by a lender of last resort; national banking systems have such a lender. The position of the Eurocurrency lenders is less clear. Announcements were made in July and September of 1974 that the various central banks would bail out commercial banks that ran into difficulty and, further, that the central banks had apportioned responsibility in cases in which international lines were crossed. There is still a question of whether existing arrangements to protect the commerical banks against a liquidity crisis caused by a massive demand for withdrawals by OPEC depositors are adequate. Even if such arrangements are not adequate, a crisis could be met by the suspension of payments. The OPEC depositors would be most unlikely to precipitate payment suspensions, except possibly in an international political crisis. Arrangements for lender-of-last-resort relationships are, of course, preferable to suspension of payments.

Similar considerations apply to the problem of potentially massive shifts from one currency to another. Existing swap arrangements could cover limited amounts for a limited period. The period is not a problem if extensions or "turnovers" are arranged. The limits on amounts are high compared to the switches of funds associated with speculation (or prudence) in anticipation of devaluations and revaluations but rather small relative to OPEC liquid funds. The question remains, however, of whether existing swap lines are adequate.

A CENTRAL BANK FOR CENTRAL BANKS

The international monetary system might become a more integral part of the capital market. Given the basic problem of maintaining the security of transnational claims, it might be advantageous for deposits with the IMF, say, to act as a buffer in the process. If commercial or near-commercial interest rates apply, why should not the IMF compete with the Eurocurrency banks and reserve-center banks for deposits from and loans to central banks and governmental and semigovernmental entities? It might utilize a slightly lower interest rate, in keeping with the limitation of both its depositors and borrowers to use central banks and official entities. In a sense, the IMF does this already but usually at rates far removed from the market. The recently developed Oil Facility came closer to market rates (7 percent). The most important consideration is whether either borrowers or lenders have reason to prefer the IMF over commercial banks, and, if so, is it legitimate to serve these parties through an international monetary agency? The most acceptable answer may not be a matter of yes or no but of how much intermediation is most appropriate. Viewed more broadly, are current banking services for the central banks of the world adequate? Once again, the question of national monetary independence arises, but it is well to consider to what extent such independence has been surrendered already, possibly to other national central banks.

If an indexed reserve unit is established by the IMF, would it not be possible for private banks to accept deposits and make loans denominated in that unit? Isn't that possible without the IMF even creating the unit? If the OPEC countries want indexed bank deposits, why are they not being offered indexed deposits?

THE "SDR LINK"

Since we are looking forward to the day when the principal component of the annual increase in world reserves would be the increase of SDRs, it has been proposed that a substantial share of newly created SDRs be issued to the less-developed countries.[10] Suppose that the annual allocation of SDRs should approach $20 billion, with dollar deficits no longer the principal increment to world reserves. A substantial portion of this—say, $7 billion a year—might be allocated to the less-developed countries, which would spend it all directly or indirectly in either the more-developed countries or in OPEC. The advantage over direct aid is that the net additional imports of the less-developed countries financed by the linked SDRs are ordinary exports to the exporting countries, not aid contributions. The true burden of the aid is indeed borne by the developed countries, but in exchange for reserves. This allocation of SDRs would permit all non-less-developed countries to run, in the aggregate, an equivalent payments surplus. As SDR allocations assumed interest charges closer to commercial rates, the advantage of link allocations to the less-developed countries would diminish. The distribution of funds in any form to the less-developed countries, say, a direct aid grant, would support the same trade patterns as with the link SDRs, but the same *payments* surpluses would not apply because the aid would be entered as an offset to the exports.

Similar in spirit to the link proposal is the proposal that the OPEC countries make extensive investments in the less-developed countries. The less-developed countries would use all or most of these funds to finance imports from the advanced countries, so that a sort of secondary trade recycling would occur following the primary recycling to the less-developed countries.[11] OPEC members have made substantial commitments, in various forms, to certain less-developed countries, and the process may go forward. While these commitments have totaled $9 billion or more,[12] the disbursements are being made slowly. To relieve extreme hardship, special additional measures are necessary in that subgroup of the less-developed countries known as the Fourth World.

SECONDARY RECYCLING IN PRACTICE

The immediate need of OECD countries for secondary recycling was reduced by the end of 1974, principally because the recession in the industrialized countries shifted the deficits to the less-developed countries as well as to the less industrialized of the developed countries (see Table 11). Italy, the paradigm case of a developed country whose political equilibrium might suffer in the absence of new provisions for secondary recycling, was removed from the danger list. She got through 1974 with her foreign reserves actually increased by $0.5 billion in spite of a record-breaking deficit on current account of about $8 billion. She borrowed over $5 billion from official sources (roughly $2 billion each from the Bundesbank and the EEC Support Fund and $1 billion from the IMF) and $3 billion through the financial markets (about $1 billion through traditional private capital flows and $2 billion through government-controlled institutions borrowing in the Euromarket). Meanwhile, her 1975 current-balance deficit is expected to fall below $2 billion,[13] which can be expected to be financed with far less trouble than 1974's $8 billion deficit. Italy's political equilibrium remains delicate, but her current-account deficit has been greatly improved—at the expense of a policy-induced recession in the growth rate, aimed both at inflation and the balance-of-payments deficit.

A major part (about $7 billion)[14] of Britain's $9 billion 1974 current-account deficit was financed by capital inflows from OPEC, while traditional capital flows and special foreign borrowing by the government and public sector provided enough to support not only the balance of the deficit but substantial official outflows.[15] Britain's 1975 deficit is expected to run well below 1974's, and, in any case, any excess over primary recycling can be covered through normal financial channels, including those forms of "compensatory borrowing" by public-sector enterprises that are now accepted as normal. Japan has swung into current-account balance in 1975:1, at least for the moment, by dint of a sharp recession and an export drive; France, by a gentler recession and export drive. These leading countries should experience no trouble in borrowing either through the Euromarket or from banks or capital markets in other leading OECD countries.

The other OECD countries are projected to run substantial deficits in 1975, particularly the "other OECD South" countries. This group is carrying the second heaviest load of deficit relative to GNP, while the Fourth World countries carry the heaviest. Some of these other OECD countries may require intergovernmental capital flows in late 1975 or 1976, but their needs are likely to be small relative to those of the Fourth World. Their access to private capital markets is better, although by no means trouble-free. The Fourth World is therefore likely to offer the most serious problems of financing current accounts. Those who, "perhaps stubbornly," maintain that "despite the remarkable resilience and adaptability of market processes, governments and international institutions must be prepared to intervene on a sizable scale to help facilitate transfers of oil payments in order to avoid a seizing up of the international economy" see "the gravest danger [lying] in the continued concentration of overall payments deficits in the less developed countries."[16]

THE FOURTH WORLD[17]

The oil price rise, bringing untold wealth to certain Third World countries, created a need for a new category of nations in international economic and political dealings: the Fourth World. That category includes all the non-oil-exporting members of the Third World.[18] For certain purposes, it must in turn be divided into two groups: the poorest and the not-so-poor. We may call the latter the developing countries. The poorest group contains 23 countries with 1972 per capita incomes below $200, averaging about $110; the developing group of 71 countries with 1972 per capita incomes above $200, averaging about $530. The poorest group offers the most difficult problems at present and limited prospects for future improvement. Almost all of the countries in this group are also on the list of 41 "most seriously affected" (MSA) countries, identified by the UN Emergency Operations as having the lowest per capita income and being hardest hit by recent world economic developments. The 23 countries in the poorest group have an aggregate population of 900 million, 55 percent of the population of the Fourth World, and 35 percent of the population of the entire non-Communist World, but with only 3.25 percent of the GNP. Their population growth rate is about 2.5 percent per year; that of the developing group about 2.8 percent. The poorest countries are located principally in Southeast Asia and in East and Central Africa. Indeed, the three countries of the Indian

developed countries would be well advised to phase out, since they are at a comparative disadvantage relative to their other products. The trade-liberalization policy should, of course, be accompanied by domestic measures to ease the transition of workers and resources out of the displaced industries.

As an alternative to commodity agreements, trade liberalization is superior on all counts. It offers lasting benefits to all concerned, while the commodity agreements are likely to be short-lived, to be costly to at least one side of the bargain, and to lead to misuse of resources. This sort of sensible readjustment cannot be made rapidly, but an organized attempt might yield strong benefits. If ever there was a textbook case of the advantage of international trade, this is it. There can be no question that if the OECD as a whole imports more from the Fourth World, the Fourth World will import more from the OECD countries. If trade liberalization proceeds on a broad front, all nations can expect to receive their share, but two sorts of difficulties should be expected. First, the Fourth World may have trouble managing its resources in order to meet the enlarged markets for their exports. Second, if the additional exports are made, there will be a strong tendency for most of the increased proceeds to go into increased imports for consumption rather than for development. While this might be a welcome development, since the peoples of the Fourth World could certainly benefit from higher levels of consumption, it means that it will be harder to finance growth through exports than through equal amounts of capital flows.

If the OECD countries manage to keep their economies close to full employment, not only will they provide better markets for the Fourth World, but they will also be more likely to give additional support to the growth of the developing countries through official capital flows. The OPEC countries can hardly be expected to increase their official capital flows to developing countries above the $10 billion level assumed in the World Bank projections. This includes about $2.5 billion in direct bilateral aid, principally to politically affiliated countries, such as from Saudi Arabia to Egypt. Another $5 billion to $6 billion is assumed to flow through multilateral agencies, princi-

pally the Oil Facility of the IMF, although the full commitment
of OPEC lenders to that Fund in 1975 has not yet been announced. If, as is generally expected, the annual OPEC surplus
on current account continues to shrink even in nominal terms
as OPEC imports rise, special effort will be required to maintain the flow of OPEC funds to the Fourth World. If maintained at the $10 billion level (in current dollars), OPEC net
official flows to the Fourth World will still come to more than 2
percent of OPEC GNP in 1980 (5 percent in 1975), as compared with 0.3 percent expected from the developed countries
in 1980 (0.38 percent in 1975). If loans to the IMF Facility are
excluded from the total as being almost commercial in nature,
the remaining OPEC official capital flows to the developing
countries will run over 2.5 percent of GNP in 1975; if maintained in nominal value, they will still be about 1 percent of
OPEC GNP in 1980.

While official capital flows from the industrial (DAC)
countries can be expected to grow in nominal terms, their rate
of growth cannot be expected to exceed the rate of inflation,
so that the most that can be hoped for is the maintenance of
the real value of DAC capital flows to the Fourth World. It
cannot be realistically hoped that either the OPEC or the
DAC countries, or both together, can come forth with a new
program that will restore the growth rate of 6 percent in
aggregate output set for the second development decade, a
rate that was actually achieved between 1960 and 1973. That
would probably require, as mentioned above, additional annual capital flows over what seems to be in prospect of $20
billion to $30 billion a year (in 1974 dollars). Funds of that
order can hardly be expected to become available. But it is
within the realm of the practical to suggest that more can be
done than is now likely to be done.

In addition to the gains that might be made from the
reduction of obstacles to exports from the Fourth World to
the OECD countries, which could possibly permit an export
increase of $12 billion a year by 1990, a greater volume of
official capital flows might be sought. In particular, for the
poorest countries of the Fourth World, the amounts of additional capital flows required to meet reasonable growth

targets are much smaller than for the developing countries. The targets are, by necessity, more modest, but the needs are more intense. While for the developing countries it would require additional flows of over $20 billion a year (in 1974 dollars) above prospective flows to finance a growth rate of 3.7 percent per year of per capita GDP instead of the 2.5 percent currently projected, for the poorest countries, an additional $2 billion a year above prospective flows would finance a growth rate of 2.0 percent per capita instead of a 0.9 percent rate. These are the estimated additional capital requirements of the respective growth rates. It should not be inferred that they would by themselves bring about the change in rates, but they would be necessary to support a program that could bring about such changes.

Certainly, the level of additional financing in the development of the poorest countries is within the realm of the practical. The IMF Oil Facility, with its interest charge close to commercial rates, is of limited use to the poorest countries, though they have used it. A UN attempt to gather matched contributions from the industrial countries and the oil exporters has come to very little because of political considerations. Total commitments made were $275 million in 1974. It was proposed by Secretary Kissinger that the IMF trust fund for the needy countries, to be financed by the sale of one-sixth of the IMF's gold, would give loans up to $2.5 billion a year for those countries. But more will be required, up to a potential total of $10 billion.[21]

A triangular arrangement to help the poorest countries of the Fourth World has much to recommend it to the industrial countries, the OPEC countries, and the Fourth World. Under such an arrangement, the OPEC countries could supply the capital to the developing countries and share the risk of default with the OECD countries and possibly also with a multilateral agency such as the World Bank. The individual OPEC countries can receive commercial or near-commercial interest on the capital invested but share jointly with the OECD countries the cost of subsidizing that interest. The Fourth World countries can get their loans on highly concessional terms made possible by the subsidies jointly provided by the OPEC and the OECD countries. They will certainly spend the proceeds of their loans in the OECD countries. The latter can then finance their deficit with OPEC in part with the proceeds of the additional exports they sell to the developing countries. The cost to them consists of the interest subsidies and the eventual losses from those guaranteed loans that are defaulted.

One such plan has been proposed to supply $3 billion a year in additional funds to the Fourth World, from 1976 to 1980.[22] The annual cost to the United States under that plan is put at $170 million, presumably not including any allowance for defaults.

The World Bank is working on the development of a "Third Window," offering development assistance on terms intermediate between the terms offered by the Bank and the concessionary terms offered by the International Development Association of the World Bank Group (IDA). It plans to offer loans at 4 to 4.5 percent below the Bank's current lending rate, with the difference subsidized voluntarily by those countries prepared to do so. Though many of the poorest countries might not be able to afford even the intermediate terms of the Third Window, it is hoped that they would benefit from more IDA resources made available to them as some of the developing countries are served by the Third Window instead of by IDA.[23]

Notes

Chapter II

[1] *Petroleum Intelligence Weekly,* June 16, 1975, p. 1.

[2] M. A. Adelman, *The World Petroleum Market* (Baltimore: The Johns Hopkins University Press, 1972), p. 140, fn. 18.

[3] Ibid., p. 134.

[4] Chase Manhattan Bank, *Capital Investment by (of) the World Petroleum Industry,* 1961 and later years. See Also Neil Jacoby, *Multinational Oil* (New York: Macmillan, 1974), p. 248. See Adelman, op. cit., p. 165, for an argument against making the above inference.

[5] Walter J. Levy, "Oil Power," *Foreign Affairs,* July 1971, p. 654.

[6] G. Henry Schuler, testimony before the Senate Foreign Relations Committee, *Petroleum Intelligence Weekly*, April 22, 1974, special supplement, p. 5.

[7] Ibid., special supplement, p. 27, and April 5, 1971, supplement; *Petroleum Economist,* October 1974, pp. 381–382.

[8] *International Petroleum Encyclopedia,* 1974, p. 11.

[9] Particularly Schuler, op. cit.

[10] Percy testimony in *Hearings before the Subcommittee on Multinational Corporations of the Senate Committee on Foreign Relations on Multinational Petroleum Companies and Foreign Policy,* 93rd Cong., January 31, February 1 and 6, 1974, part 5, p. 212.

[11] Robert Mabro, "Can OPEC Hold the Line," *Middle East Economic Survey,* February 28, 1975, supplement; and an anonymous OPEC official quoted in *Petroleum Intelligence Weekly*, March 10, 1975.

[12] Levy, op. cit., p. 655.

[13] OECD, *Energy Prospects to 1985*, vol. II, p. 115. The source gives the price as $9 in 1972, taken to be equivalent to $10.50 in 1974. OPEC 1980 sales are given as 27.6 m.b.d. in source, which apparently is in error.

[14] Federal Energy Administration (FEA), *Project Independence Report* (Washington, D.C.: Project Independence, November 1974), Table VII-1, p. 355.

[15] *New York Times,* February 8, 1975.

[16]M. A. Adelman, testimony before the Subcommittee on Multinational Corporations of the Committee on Foreign Relations, U.S. Senate, January 29, 1975.

[17]FEA, op. cit., p. 367. This estimate is made in the context of the evaluation of an embargo, but it should carry over to the case of a self-inflicted cutback.

[18]An additional allowance for the cost of oil production should be deducted from the future cost of the substitute, but that would not significantly affect the examples given.

[19]*New York Times,* April 16, 1973.

[20]*The Wall Street Journal,* December 13, 1973; *Petroleum Intelligence Weekly,* December 17, 1973, p. 3.

[21]Statement by Thomas O. Enders, Assistant Secretary of State for Economics and Business Affairs, *Department of State Bulletin,* March 10, 1975, p. 307.

[22]*New York Times,* December 17, 1974.

[23]Ibid., September 2, 1975.

[24]Ibid., September 12, 1975.

[25]*New York Times Magazine,* September 14, 1975.

Chapter III

[1]Shell Chairman McFadyean, quoted in *Middle East Economic Survey,* May 23, 1975, p. 4.

[2]*BP Statistical Review of the World Oil Industry,* 1974, p. 16.

[3]George L. Perry, in forthcoming publication of the Brookings Institution, edited by Edward R. Fried and Charles Schultze.

[4]It also excludes the price rises of producers' goods that have not yet been embodied in consumers' goods.

[5]Perry, op. cit.

[6]Ibid., Table 2-8.

[7]Ibid., Table 2-9.

[8]Jon McLin, *Oil, Money and the Common Market,* vol. IX, no. 2 (New York: American Universities Field Staff, July 1974).

[9]OECD, *Economic Outlook,* December 1974, pp. 61–62

[10]Ibid., July 1975, p. 57.

[11]A practical problem arises in this case in that much of the primary recycling goes to institutions that are not part of any particular country. Deposits in Eurocurrency banks, loans to the IMF and World Bank fall into this class. For simplicity in the text, we shall assume that all primary recycling can be identified as going to specified countries. From a practical point of view, capital transfers from "extranational" entities may be regarded as primary recycling in the present context, as if they short-circuited the extranational middleman.

[12]This may be demonstrated as follows:

Let CB = the country's current balance with all other countries

D = the country's deficit with OPEC countries

S = the country's surplus with other non-OPEC countries

R=primary recycling received from OPEC countries; that is, its current-account surplus with non-OPEC countries will equal its overall deficit with OPEC countries.

By the rules of current-balance accounting:
$$CB = S - D$$

so if
$$CB = - R, \text{ then } - R = S - D, \text{ or } S = D - R$$

Chapter IV

[1] *Newsweek*, February 3, 1975, p. 27.

[2] Jahangir Amuzegar of Iran, quoted in *New York Times*, April 26, 1974.

[3] Otmar Emminger, "International Financial Markets and the Recycling of Petrodollars," *The World Today*, March, 1975, p. 97.

[4] See Franco Modigliani and Hossein Askar: *The Reform of the International Payments System, Essays in International Finance*, no. 89, September 1971.

[5] Sanford Rose, "The Misguided Furor about Investment from Abroad," *Fortune*, May 1975, p. 172.

[6] Charles W. Robinson, *Department of State Bulletin*, vol. LXXII, no. 1865, March 24, 1975, p. 379.

Chapter V

[1] See Hollis Chenery, "Restructuring the World Economy," *Foreign Affairs*, January 1975, for a scenario of this sort.

[2] Thomas O. Enders statement to the Subcommittee on Multinational Corporations of the Senate Committee on Foreign Relations, February 14, 1975. *Department of State Bulletin*, March 10, 1975, p. 316.

[3] *IMF Survey*, April 28, 1975, p. 104.

[4] Ibid., September 15, 1975, p. 261.

[5] Ibid., April 28, 1975.

[6] Ceteris Paribus, pseud., "Recycling," *Foreign Policy*, no. 17 Winter 1974–75, pp. 85–87. The author is reported to be a "high Treasury official."

[7] Peter Kenen, "Reforming the Monetary System—You Can't Get There from Here," *Euromoney*, October 1974.

[8] "Report to the Board of Governors by Committee of Twenty," *International Monetary Reform* (Washington, D.C.: International Monetary Fund, 1974).

[9] IMF, *Annual Report*, 1974, p. 127.

[10] See Lord Kahn, "SDR's and AID," *Lloyd's Bank Review*, October 1973, pp. 1–18. See also Y. S. Park, *The Link between Special Drawing Rights and Development Finance*, essays in international finance, no. 100. (Princeton, N.J.: Princeton University Press, September 1973).

[11]The "Carli Plan" runs along these lines. See *Business Week*, October 2, 1974.

[12]World Bank, *Annual Report*, 1975, p. 42.

[13]OECD, *Economic Outlook*, July 1975, p. 77.

[14]Bank of England, *Quarterly Bulletin*, reported in *Economist*, September 20, 1975, p. 84.

[15]OECD, op. cit.

[16]Robert V. Roosa and Keith B. Josephson, "Oil-Caused Payments Problems Still Plague World Economy," *American Banker*, sect. 2, August 29, 1975, pp. 1A and 24A.

[17]The projections in this section are based on those of the World Bank staff in their reports of July 8, 1974, and April 8, 1975.

[18]Some sources reserve the term "Fourth World" for the group of countries here denominated the "poorest group."

[19]This refers to international financial developments only. Domestically, some of these countries had disastrous crop years.

[20]Estimate of World Bank staff.

[21]*IMF Survey*, September 15, 1975.

[22]Trilateral Task Force on Relations with Developing Countries, *OPEC, the Trilateral World and Developing Countries: New Arrangements for Cooperation 1976–80*, New York, 1975. Plan sketched in *Trialogue*, no. 6, Winter 1974–75.

[23]World Bank, *Annual Report*, 1975, p. 14.